GENETICS – RESEARCH AND ISSUES

ACETYLCHOLINE RECEPTORS IN HEALTH AND DISEASE

GENETICS – RESEARCH AND ISSUES

Additional books and e-books in this series can be found on Nova's website under the Series tab.

GENETICS – RESEARCH AND ISSUES

ACETYLCHOLINE RECEPTORS IN HEALTH AND DISEASE

ADELAIS EROS GUPTA
EDITOR

NOTICE TO THE READER

The Publisher has taken reasonable care in the preparation of this book, but makes no expressed or implied warranty of any kind and assumes no responsibility for any errors or omissions. No liability is assumed for incidental or consequential damages in connection with or arising out of information contained in this book. The Publisher shall not be liable for any special, consequential, or exemplary damages resulting, in whole or in part, from the readers' use of, or reliance upon, this material. Any parts of this book based on government reports are so indicated and copyright is claimed for those parts to the extent applicable to compilations of such works.

Independent verification should be sought for any data, advice or recommendations contained in this book. In addition, no responsibility is assumed by the Publisher for any injury and/or damage to persons or property arising from any methods, products, instructions, ideas or otherwise contained in this publication.

This publication is designed to provide accurate and authoritative information with regard to the subject matter covered herein. It is sold with the clear understanding that the Publisher is not engaged in rendering legal or any other professional services. If legal or any other expert assistance is required, the services of a competent person should be sought. FROM A DECLARATION OF PARTICIPANTS JOINTLY ADOPTED BY A COMMITTEE OF THE AMERICAN BAR ASSOCIATION AND A COMMITTEE OF PUBLISHERS.

Additional color graphics may be available in the e-book version of this book.

Library of Congress Cataloging-in-Publication Data

ISBN: 978-1-53615-447-4

Published by Nova Science Publishers, Inc. † New York

CONTENTS

PREFACE

Acetylcholine Receptors in Health and Disease opens with a review of the results of an investigation on the cholinergic modulation of excitatory synaptic transmission in the frog tectum carried out in the laboratory of neurophysiology at the Lithuanian University of Health Sciences. Experiments were done *in vivo* on the common grass frog *Rana temporaria*.

Next, the authors review the function of M4 MR and discuss possible detection methods. M4 MR regulated locomotion is studied in conjunction with recent data on consequences to biorhythms.

In the closing chapter, the authors review the environmental enrichment paradigm in rodents, as well as their effects on neurobiological, physiological and behavioral variables in preclinical studies.

Chapter 1 - The chapter reviews results of investigation of the cholinergic modulation of excitatory synaptic transmission in the frog tectum gained in Laboratory of Neurophysiology, Lithuanian University of Health Sciences. Experiments were done *in vivo* on common grass frogs *Rana temporaria*. A certain level of the endogenous acetylcholine exists in the frog tectum at normal physiological conditions due to a persistent activity of cholinergic nuclei of the frog brain, primarily the nucleus isthmi. The authors have demonstrated that this background acetylcholine activates presynaptic nicotinic receptors, causing the steady potentiation of the retinotectal synaptic transmission up to 1.7 times. The authors have called this type of nicotinic potentiation tonic nicotinic potentiation and corresponding presynaptic receptors – tonic nicotinic receptors. Results of the experiments have demonstrated that not only glutamate as a main mediator but also acetylcholine as a co-mediator release from the retinotectal synapses during firing of the retinotectal fiber. This co-released acetylcholine adds to background acetylcholine increasing the extracellular concentration of the acetylcholine. During a relatively strong burst of the retinotecal fiber (4 and more action potentials with interpulse intervals of 10 ms) the concentration of the acetylcholine increases sufficiently to activate another kind of presynaptic nicotinic receptors, called phasic nicotinic receptors. Activation of these receptors leads to a transient increase of the retinotectal synaptic transmission. This kind of potentiation of the retinotectal synaptic transmission the authors have called phasic nicotinic potentiation. Magnitude of the phasic nicotinic potentiation depends on the burst strength and ranges from 1.4 to 2.2 times. It also depends on the time after the burst. The phasic nicotinic potentiation subsides within a period of ~1 min after the end of the burst. Results of the experiments have demonstrated that the phasic and tonic nicotinic receptors exhibit different affinity to the antagonist d-tubocurarine. The tonic receptors were of higher affinity to the d-tubocurarine than the phasic receptors. This result the authors have extrapolated to the agonist acetylcholine, too, suggesting that the tonic nicotinic receptors are of higher affinity to the acetylcholine than the phasic receptors. Such a difference in the features would conform well to different functions of the receptors. Results of the experiments have demonstrated that not only the retinotectal synaptic transmission but also the tectum intrinsic recurrent excitatory synaptic transmission (positive feedback) is modulated by the acetylcholine, although in opposite direction than the retinotectal synaptic transmission. The acetylcholine co-released during a burst of the retinotectal fiber activates muscarinic receptors located (presynaptically and/or postsynaptically) in the recurrent synapses made up by the pear-shaped neurones. The activation of the muscarinic receptors causes a depression (not potentiation) of the recurrent excitatory potentials up to 2 times. This muscarinic inhibition of the recurrent excitation occurs with a delay of ~80 ms.

Chapter 2 - Muscarinic receptors belong to the class of G protein-coupled receptors and exist in five subtypes (M_1-M_5). M_4 muscarinic receptors (M_4 MR) are those that are together with M_2 MR coupled to G_i protein and thus inhibit adenylyl cyclase. Even-numbered muscarinic receptors are considered primarily as pre-synaptic receptors. However, both M_2 MR and M_4 MR are localized both pre-synaptically and post-synaptically. As cholinergic autoreceptors, M_2 and M_4 provide feedback control of acetylcholine release. M_4 MR are coupled to $G_{i/o}$ G proteins and to G_s, which provides further level of signal fine tuning. M_4 MR are spontaneously active and can cause constitutive inhibition of adenylyl cyclase. M_4 MR were also shown to enhance Ca^{2+} currents via the effect on Ca_{V1} channels. Sequencing study on rat M_4 MR has shown the presence of cell type-specific silencer element in the promoter region and that expression of M_4 MR is regulated by the neuron-restrictive silencer element/repressor element 1. All MR reveal relatively high amino acid sequence similarity and thus only a few MR ligands are selective. This makes the detection of MR in respective tissue sometimes problematic. Typically, it is not possible to use specific radioligands. Detection of protein – using western blot or immunohistochemical methods – suffers from unexpectedly high non-specificity of antibodies as verified on knockout mice. Similarly, the evaluation of gene expression is also not an appropriate method for M_4 MR detection as the mRNA level usually does not correspond with receptor levels. M_4 MR have been associated with various organism functions during the past years. Initially, the role of MR could be elucidated only by means of pharmacological studies. These studies have indicated the important role of MR in several brain processes including learning and memory, attention, locomotion, thermoregulation, sleep and wakefulness, food intake and reward. Here, the author will review the function of M_4 MR. Next, possible detection methods (radioligand binding, autoradiography, western blotting, gene expression) will be discussed with special interest to knockout mice as specific M_4 MR function model. Next, M_4 MR regulated locomotion will be reviewed together with the very recent data on consequences to biorhythms. To sum up, it is important to remember that M_4 MR are a subtype of MR with various functions and connection to biorhythms.

Chapter 3 - Nicotine and agonists of alpha7 nicotinic acetylcholine receptors (nAChRs) are being evaluated as a possible treatment for cognitive symptoms associated to schizophrenia, Alzheimer's disease and other neuropscyhiatric and neurodegenerative disorders. It has been proposed that these drugs can act as cognitive enhancers, improving memory and learning in different tasks. Therefore, it could be of interest to evaluate their effects in preclinical studies in which different drugs are administered in conjunction with other therapeutic strategies such as environmental interventions based on complex environments including higher social, cognitive and physical stimulation than standard environments. Prior studies in rodents suggest that these enriched environments induce behavioral and neurobiological changes which can be related to diminished anxiety-like behavior and an improvement in the performance of learning and memory tasks. In the present chapter, the authors will review the environmental enrichment (EE) paradigm in rodents and their effects on neurobiological, physiological and behavioral variables in preclinical studies. The authors will also present results obtained in their laboratory regarding effects of nicotine and PNU 282987 in interaction with EE in different behavioral tasks. The authors will conclude indicating the implications that experimental results may have for the search and identification of new therapeutic approaches aimed to counteract or delay cognitive deficits observed in neurodegenerative or neuropsychiatric disorders.

In: Acetylcholine Receptors …
Editor: Adelais Eros Gupta

ISBN: 978-1-53615-447-4

Chapter 1

CHOLINERGIC MODULATION OF EXCITATORY SYNAPTIC TRANSMISSION IN THE FROG TECTUM

***A. Baginskas*[1,*], *A. Kuras*[2] and *A. Grigaliūnas*[1]**

[1]Department of Physics, Mathematics and Biophysics, Medical Academy, Lithuanian University of Health Sciences, Kaunas, Lithuania

[2]Laboratory of Neurophysiology, Neuroscience Institute, Medical Academy, Lithuanian University of Health Sciences, Kaunas, Lithuania

ABSTRACT

The chapter reviews results of investigation of the cholinergic modulation of excitatory synaptic transmission in the frog tectum gained in Laboratory of Neurophysiology, Lithuanian University of Health Sciences. Experiments were done *in vivo* on common grass frogs *Rana temporaria*.

A certain level of the endogenous acetylcholine exists in the frog tectum at normal physiological conditions due to a persistent activity of

[*] Corresponding Author's E-mail: armuntas.baginskas@lsmuni.lt.

cholinergic nuclei of the frog brain, primarily the nucleus isthmi. We have demonstrated that this background acetylcholine activates presynaptic nicotinic receptors, causing the steady potentiation of the retinotectal synaptic transmission up to 1.7 times. We have called this type of nicotinic potentiation tonic nicotinic potentiation and corresponding presynaptic receptors – tonic nicotinic receptors.

Results of the experiments have demonstrated that not only glutamate as a main mediator but also acetylcholine as a co-mediator release from the retinotectal synapses during firing of the retinotectal fiber. This co-released acetylcholine adds to background acetylcholine increasing the extracellular concentration of the acetylcholine. During a relatively strong burst of the retinotecal fiber (4 and more action potentials with interpulse intervals of 10 ms) the concentration of the acetylcholine increases sufficiently to activate another kind of presynaptic nicotinic receptors, called phasic nicotinic receptors. Activation of these receptors leads to a transient increase of the retinotectal synaptic transmission. This kind of potentiation of the retinotectal synaptic transmission we have called phasic nicotinic potentiation. Magnitude of the phasic nicotinic potentiation depends on the burst strength and ranges from 1.4 to 2.2 times. It also depends on the time after the burst. The phasic nicotinic potentiation subsides within a period of ~1 min after the end of the burst.

Results of the experiments have demonstrated that the phasic and tonic nicotinic receptors exhibit different affinity to the antagonist d-tubocurarine. The tonic receptors were of higher affinity to the d-tubocurarine than the phasic receptors. This result we have extrapolated to the agonist acetylcholine, too, suggesting that the tonic nicotinic receptors are of higher affinity to the acetylcholine than the phasic receptors. Such a difference in the features would conform well to different functions of the receptors.

Results of the experiments have demonstrated that not only the retinotectal synaptic transmission but also the tectum intrinsic recurrent excitatory synaptic transmission (positive feedback) is modulated by the acetylcholine, although in opposite direction than the retinotectal synaptic transmission. The acetylcholine co-released during a burst of the retinotectal fiber activates muscarinic receptors located (presynaptically and/or postsynaptically) in the recurrent synapses made up by the pear-shaped neurones. The activation of the muscarinic receptors causes a depression (not potentiation) of the recurrent excitatory potentials up to 2 times. This muscarinic inhibition of the recurrent excitation occurs with a delay of ~80 ms.

Keywords: cholinergic modulation, nicotinic, muscarinic, acetylcholine receptors, co-release, co-mediator, neuromodulation, retinotectal synaptic transmission, recurrent excitation, feedback, frog, tectum column

INTRODUCTION

Experimental results evidence that acetylcholine modulates activity of neural circuits of the brain by changing strength of synaptic transmission (see, for example, Obermayer et al., 2017; Piccioto et al., 2012). Results of our experiments have demonstrated the cholinergic modulation of synaptic transmission in the frog tectum column. Here we present a review of those results.

Tectum is a main center of the frog brain for processing of the optical information. Ganglion cells of the retina of the frog eye send their axons, called retinotectal or optic fibers, to the tectum where they make up synaptic connections with the tectum neurons. Axon of a single ganglion cell (individual retinotectal fiber) makes up connections with a group of tectum neurons that compose a functional unit (elementary neuronal network) of the tectum, called tectum column. The results of morphological (Szekely and Lazar, 1976; Lara et al., 1982) and our electrophysiological (Kuras et al., 2007; Baginskas and Kuras, 2008) studies have shown that the tectum column consists of five recurrent pear-shaped neurons, one efferent pyramidal neuron, and two inhibitory interneurons (stellate cell and amacrine cell) (Figure 1).

The retinotectal (optic) fibres project largely to the dendrites of recurrent pear-shaped neurons, making up axodendritic glutamatergic synapses (Figure 1, ***b***). The minor projection is to the dendrites of pyramidal cell (Figure 1, ***a***). The pear-shaped neurons form intrinsic recurrent excitatory loops (positive feedback) by making up axodendritic (Figure 1, ***c***) and dendrodendritic (Figure 1, ***d***) glutamatergic synapses with the pyramidal cell. The pear-shaped neurones also make up axodendritic synapses with the inhibitory neurons (Figure 1, ***e***). The inhibitory neurons make up somatodendritic inhibitory synapses with the dendrites of excitatory pear-shaped neurons (Figure 1, ***f***).

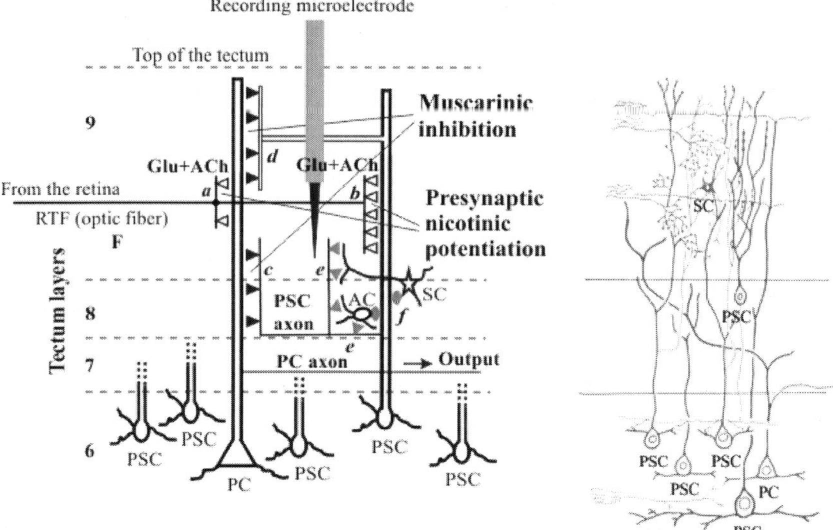

Figure 1. Scheme of the frog tectum column. The right side of the diagram shows a drawing from the Figure 11 of the article by Szekely and Lazar (1976). RTF marks retinotectal (optic) fiber, PSC – pear-shaped cells (recurrent neurons of the column), PC – efferent pyramidal cell (output neuron of the column). Axons of the PSC and PC neurons originate from proximal dendritic stems, not from somas. AC and SC are amacrine and stellate cells (inhibitory neurons of the column). Retinotectal fiber makes up major excitatory synaptic input to the dendrites of pear-shaped neurons (*b*) and minor excitatory synaptic input to the dendrites of pyramidal cell (*a*). Pear-shaped neurones make up axodendritic (*c*) and dendrodendritic (*d*) excitatory synapses with the dendrites of pyramidal cell. They also make up axodendritic and axosomatic synapses with the inhibitory neurones (*e*). The inhibitory neurons make up somatodendritic inhibitory synapses with the dendrites of pear-shaped neurons (*f*).

Results of the experiments reviewed below address three questions:

1. Is there cholinergic modulation of the retinotectal transmission (synapses ***a*** and ***b*** in Figure 1)?
2. Is there cholinergic modulation of the intrinsic recurrent excitatory transmission (synapses ***c*** and ***d*** in Figure 1)?

3. Does acetylcholine release as a co-mediator from the glutamatergic retinotectal terminals (synapses ***a*** and ***b*** in Figure 1) during a firing of the retinotectal fiber?

The experimental procedure that lets to stimulate a single ganglion cell and record the evoked activity of tectum column is described below.

METHOD

Frog Surgery

Experiments were done *in vivo* on adult common grass frogs *Rana temporaria*. The experiments were carried out in accordance with the "Principles of laboratory animal care" (NIH publication No. 86-23 revised in 1985) as well as with the European Communities Council Directive of 24 November 1986 (86/609/EEC) and were approved by the Animal Care and Use Committee of the State Food and Veterinary Service of Lithuania (No. 0167).

During the surgical manipulations, frogs were generally anesthetized with a high concentration of CO_2 (Forslid et al., 1986; King et al., 1999). This method of anaesthesia was chosen due to a brief period of recovery from the anaesthesia (Forslid et al., 1986). The regions above the tectum and around the eye that were operated on were additionally locally numbed by a subcutaneous injection of the analgesic procaine. The dorsal tectum was exposed in the way Maturana et al. (1960) have described: The skin above the tectum was removed and the skull was trepanned. *Dura mater* was excised and *pia mater* was removed using a tungsten wire sharpened to a 1–2-µm diameter under the microscope control. The contralateral to the opened tectum eye was prepared according to the method of George and Marks (1974): The upper eyelid, the nictitating membrane, and the sclera were excised. The lens and hyaloid were removed by suction. The eyeball cavity was filled with the Ringer's solution (in mM: 116 NaCl, 2.5 KCl, 1.8 $CaCl_2$, 1.0 $MgCl_2$, 1.2 $NaHCO_3$, and 0.17 $NaH_2PO_4 \cdot 2H_2O$;

pH 7.3–7.4). The exposed dorsal tectum was perfused with the same Ringer's solution. Then the frogs were immobilized with an intramuscular injection of 0.2–0.3 mg of d-tubocurarine, intubated and ventilated by a mechanical ventilator with a tidal volume of 2–3 ml and frequency of 6–10 breaths per minute. Subsequent injections of 0.1 mg of d-tubocurarine were delivered every 20–30 min to keep the frog immobilized during the experiments. All recordings were done in the dark at the ambient temperatures of 17–21^{0}C. At the end of the experiments the animals were killed with an anaesthetic overdose.

Stimulation and Recording

The method of stimulation of a single retina ganglion cell or its axon and recording of the evoked response in the tectum was established and described in Kuras and Gutmaniene (1997, 2001). Figure 2 illustrates the *in vivo* experiment.

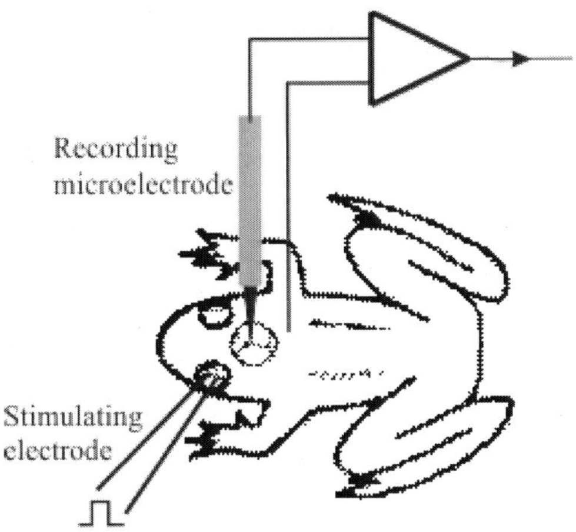

Figure 2. Illustration of the experiment. Electrical pulses were delivered to the frog eye's retina through the multichannel stimulating electrode. Excitation of a single retina ganglion cell was achieved. Evoked activity of the tectum column was recorded by the carbon-fiber microelectrode, inserted into a half of the tectum that is contralateral to the eye operated on.

Multichannel stimulating electrode consisted of two 5-channel sub-blocks. The channels were made of tungsten wires of a diameter of 40 μm with a distance between the adjacent wires (channels) of 70–250 μm. Direct current resistance of a channel was equal to 180–190 kΩ. The stimulating electrode was brought into the eye cavity and gently pushed approaching the retina. Usually the electrode was targeted to the nasoventral quadrant of the retina.

Carbon-fibre microelectrode of improved signal-to-noise ratio (Kuras and Gutmaniene, 1995) and length of the bare tip of 45–70 μm was used to record responses from the tectum. The recording microelectrode was inserted into contralateral half of the tectum and moved to the depth of F layer of the tectum (see Figure 1). F layer lies 200–310 μm deep from the surface of the tectum (Chung et al., 1974; Kuras and Khusainoviene, 1986).

Electrical pulses were delivered to the retina through different channels of the stimulating electrode until the response at the site of the insertion of the microelectrode was found. Then, a unitary response was achieved ("unitary response" meaning a response evoked in the tectum by firing of a single retina ganglion cell or it's axon) by decreasing an amplitude of the current pulse. Unitary characteristic of the response was checked following generally accepted criteria for "all-or-none" events (George and Marks, 1974; Luscher et al., 1983; Stern et al., 1992; Raastad, 1995; Kuras and Gutmaniene, 1997, 2001): 1) presence of the response in a half of stimulation trials, when the stimulus strength is of threshold value; 2) sudden disappearance of the response in all of the stimulation trials, when the stimulus strength gets below the threshold value; 3) presence of the response in all of stimulation trials, when the stimulus strength is slightly above the threshold value; 4) near doubling of amplitude of the response with a further increase of the stimulation strength.

The current pulses, which elicited unitary responses, were of duration of 50 μs and had an amplitude in the range of 8–61 μA. Stimulation of a single retina ganglion cell or its axon was considered reliable, if an increase of the stimulation strength by 4–6 μA had no effect on the response and a decrease by 0.2–0.8 μA completely abolished it.

Not only single current pulses, but also bursts of 2–10 pulses with interpulse intervals of 10–25 ms have been delivered to the retina. Time interval between the successive bursts was set to be of 3 minutes or longer to give the time for the tectum column to return to the resting state.

The F layer location of the recorded responses was confirmed not only by the depth of the tip of the recording microelectrode being of 200–310 μm from the surface of the tectum, but also by the latency of the responses with respect to the stimulus artifact being of 5.6–10.7 ms (7.88 ± 0.12 ms on average).

A presence of responses in other layers of the tectum was checked by moving the recording electrode up and down. We have never found the response in layers of unmyelinated retinotectal axons. Occasionally we have found the response in G layer. In such a case, the circumstances of the stimulation (channel pairs, position of the stimulating electrode) were changed until genuine unitary response was recorded from the layer F. Strength of the threshold stimulus was monitored periodically throughout the experiment.

Application of Drugs

The solutions of specific nicotinic acetylcholine receptor antagonist d-tubocurarine (D-TC, 10, 50 μM), nonspecific acetylcholine receptor agonist carbamylcholine chloride (CCh, 100 μM), specific muscarinic acetylcholine receptor agonist oxotremorine-M (Ox-M, 10 μM), nonspecific glutamate receptor antagonist kynurenic acid (KYNA, 0.6 mM), AMPA/kainate glutamate receptor antagonist CNQX (5 μM), were applied onto the surface of the tectum by the perfusion at a rate of 0.4 ml/min. The solutions of the drugs were prepared just before the use. The chemicals were purchased from Sigma–Aldrich Co.

RESPONSES OF THE FROG TECTUM COLUMN

Responses of the frog tectum column elicited by firing of a single retina ganglion cell or its axon and recorded by the microelectrode inserted into F layer of the tectum (see Figure 1) are described in detail in Chapter 3 of Nova's book "Frogs Genetic Diversity, Neural Development and Ecological Implications" (Baginskas and Kuras, 2014). The responses are shown in Figure 3.

Features of the responses depend on firing intensity of the retina ganglion cell. Number and frequency of spikes quantifies intensity of the discharge.

Response of the tectum column to a single action potential of an individual retinotectal (optic) fiber consisted of unitary action potential (AP) and unitary fast synaptic potential (fSP). Average amplitude of the APs was equal to 87 ± 4 μV, and the average amplitude of the fSPs, A_{fSP}, was equal to 134 ± 5 μV. The unitary action potential reflects spikes in the presynaptic terminals of the retinotectal fiber (***a*** and ***b*** terminals in Figure 1). The unitary fast synaptic potential reflects excitatory glutamatergic postsynaptic potentials in the synapses made up by the retinotectal fiber with the neurons of the tectum column (***a*** and ***b*** synapses in Figure 1) (Gutman and Kuras, 1974; Kuras and Khusainoviene, 1986; Kuras and Gutmaniene, 1995, 1997).

Response of the tectum column to a burst of moderate intensity of 2–4 action potentials with interpulse intervals of 10–25 ms displayed a slow negative wave (sNW) following the fSPs, and recurrent synaptic potentials (rSPs) superimposed on the wave. Average magnitude and duration of the sNW were equal to 74 ± 6 μV and 137 ± 8 ms, respectively. Average amplitude of the recurrent action potentials (rAPs) and recurrent synaptic potentials, A_{rSP}, were equal to 45 ± 1 μV and 42 ± 1 μV, respectively. The sNW reflects an activation of persistent L-type calcium currents in the dendrites of pear-shaped neurones of the tectum column, which lead to a steady state depolarization of the dendrites and persistent firing of the pear-shaped neurones (Baginskas and Kuras, 2008).

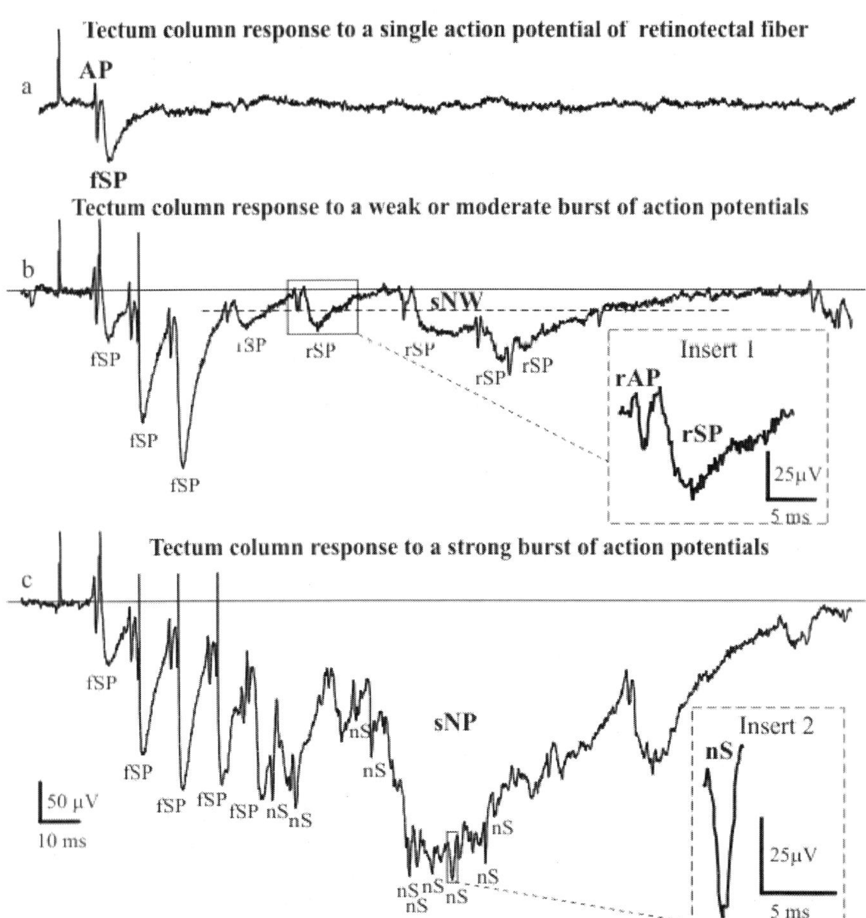

Figure 3. Responses of the tectum column to the bursts of action potentials of an individual retinotectal (optic) fiber. The features of the responses depended on the burst intensity. (a) Single retinotectal action potential (AP) evokes in the tectum fast excitatory synaptic potential (fSP). (b) Burst of three action potentials with interpulse intervals of 10 ms evokes slow negative wave (sNW) following the fSPs with the recurrent excitatory synaptic potentials (rSPs) superimposed on the wave. Insert 1 shows the recurrent action potential (rAP) and associated with it recurrent synaptic potential (rSP). (c) Burst of five action potentials with interpulse intervals of 10 ms evokes slow negative potential (sNP) following the fSPs with the negative (or negative-positive) spikes (nS) superimposed on the wave. Insert 2 shows the negative spike.

This in turn activates the recurrent dendrodendritic (***d*** synapses in Figure 1) and axodendritic (***c*** synapses in Figure 1) excitatory synapses, which is reflected in the recordings by a presence of the rAPs and rSPs riding of the back of the sNW (Kuras et al., 2004, 2006, 2007; Baginskas and Kuras, 2008).

Response of the tectum column to a burst of high intensity of 5 or more action potentials with interpulse intervals of 10–25 ms displayed a slow negative potential (sNP) following the fSPs, and negative (or negative-positive) spikes (nS) superimposed on the sNP. Average magnitude and duration of the sNP were equal to 315 ± 13 μV and 141 ± 5 ms, respectively. Average amplitude of the negative (negative-positive) spikes was equal to 39 ± 1 μV. The sNP reflects an activation of the NMDA receptors in the membrane of efferent pyramidal neuron of the tectum column (**PC** neuron in Figure 1). The activation of the NMDA receptors leads to a firing of efferent pyramidal neuron, which is reflected in the recordings by a presence of the negative (negative-positive) spikes riding on the back of the sNP (Kuras et al., 2006, 2007; Baginskas and Kuras, 2009).

From what is said above follows, that effects of the acetylcholine and other endogenous and exogenous substances on the retinotectal excitatory synaptic transmission and intrinsic tectal recurrent excitatory synaptic transmission can be investigated by measuring amplitudes and other characteristics of the fSPs and the rSPs seen in the recordings of the activity of the tectum column.

Nicotinic Potentiation and Muscarinic Depression

Results of the studies (Kuras and Gutmaniene, 2001; Baginskas and Kuras, 2011) have demonstrated existence of cholinergic modulation of synaptic transmission in the frog tectum.

The nonspecific agonist of the acetylcholine receptors carbamylcholine (CCh) has been applied to the frog brain during the experiments.

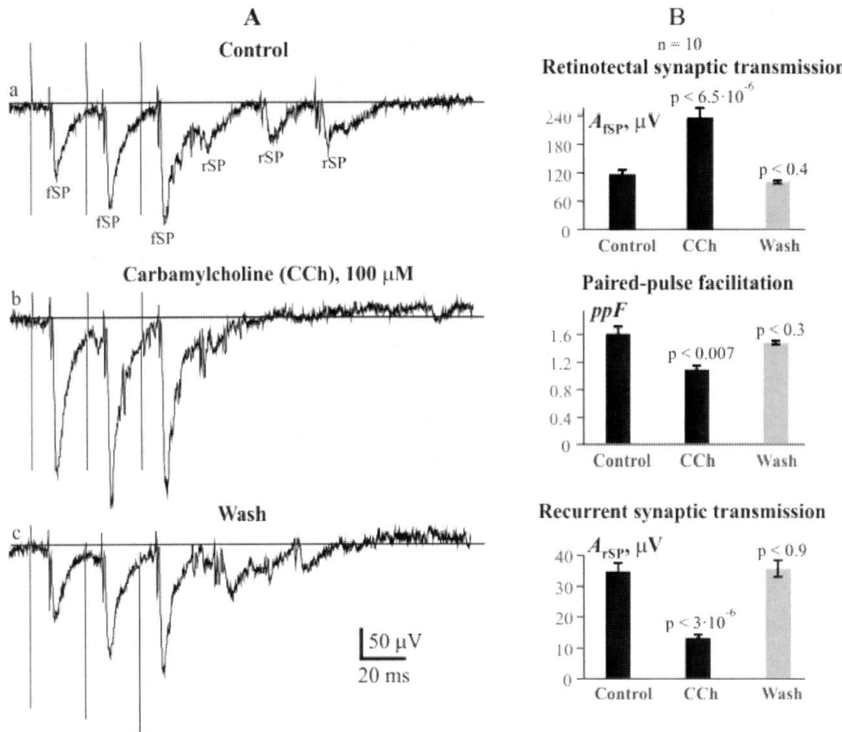

Figure 4. Effect of the nonspecific acetylcholine receptor agonist carbamylcholine chloride (CCh) on the retinotectal and recurrent synaptic transmission in the frog tectum. A) The plot shows tectal responses to electrical stimulation of a single retinal ganglion cell or its axon with a train of three pulses at interpulse intervals of 25 ms, recorded in control, CCh (100 μM) and washing conditions. The burst of three action potentials of the ganglion cell elicits in the tectum the fast synaptic potentials (fSPs) and slow negative wave with the recurrent synaptic potentials (rSPs) riding on the back of the wave. The CCh (100 μM) greatly enhances the fast synaptic potentials (fSPs), generated by the retinotectal synaptic transmission, and depresses the recurrent synaptic potentials (rSPs), generated by the intrinsic recurrent excitatory synaptic transmission. This effect of CCh is reversible. Paired-pulse facilitation (*ppF*) is estimated as a ratio of the amplitudes of the second and first fSPs in the burst. B) Average of 10 experiments. A_{fSP} – average amplitude of the fast synaptic potentials, A_{rSP} – average amplitude of the recurrent synaptic potentials, *ppF* – paired pulse facilitation.

The effects of CCh (100 μM) on the retinotectal and recurrent excitatory synaptic transmission have been investigated by measuring the amplitude and paired-pulse facilitation of the fSPs and the amplitude of the rSPs. The paired-pulse facilitation (*ppF*) of the retinotectal transmission is defined as a ratio of amplitudes of the second and first fSPs in the burst, i.e., $ppF = A_{fSP2}/A_{fSP1}$. Results of the experiments are presented in Figure 4. Perfusion of the tectum surface with 100 μM solution of CCh led to a considerable increase of the retinotectal synaptic transmission and decrease of the intrinsic recurrent excitatory synaptic transmission. Amplitude of the fSPs increased from the value of 113 ± 12 μV to the value of 233 ± 23 μV, i.e., 2.1 ± 0.1 times. Amplitude of the rSPs decreased from the value of 34.5 ± 2.9 μV to the value of 13 ± 1.1 μV, i.e., 2.7 ± 0.14 times. The paired-pulse facilitation (*ppF*) of the fSPs decreased from the value of 1.6 ± 0.11 to the value of 1.08 ± 0.06, indicating the presynaptic site of action of the CCH on the fSPs. (Measurements of paired pulse facilitation are used to determine the presynaptic (or non-presynaptic) origin of the effect. If an effect or phenomenon is accompanied by a change in the paired pulse facilitation, than the effect or phenomenon is considered to be of presynaptic origin (Parnas and Segel, 1982; Jia et al., 2008; Yamamoto et al., 2010; Walz et al., 2010; Kealy and Commins, 2010)). Thus the acetylcholine receptors, responsible for the potentiation of the retinotectal transmission, are located on the presynaptic terminals of the retinotectal fibers (***a***, ***b*** terminals in Figure 1).

Acetylcholine receptors, responsible for the depression of the recurrent excitatory transmission, could be located on the presynaptic terminals of the axons of pear-shaped neurones (***c***, ***d*** terminals in Figure 1) and/or in the membrane of the postsynaptic pyramidal cell (**PC** neuron in Figure 1). The effect of the CCh was reversible.

Opposite effect of the CCh on the recurrent synaptic transmission comparing to the retinotectal synaptic transmission could be attributed to the different types of acetylcholine receptors – nicotinic and muscarinic. This proposition has been tested experimentally via application of the specific muscarinic acetylcholine receptor agonist oxotremorine-M (Ox-M). Results of the experiments are shown in Figure 5.

Perfusion of the tectum surface with 10 μM solution of Ox-M had no effect on the retinotectal synaptic transmission and substantially decreased the intrinsic recurrent excitatory synaptic transmission. Amplitude of the fSPs in the control and Ox-M conditions was equal to 125 ± 10 μV and 127 ± 10 μV, respectively. Amplitude of the rSPs decreased from the value of 39 ± 3 μV to the value of 18 ± 1 μV, i.e., 2.2 ± 0.1 times. The effect of Ox-M was reversible.

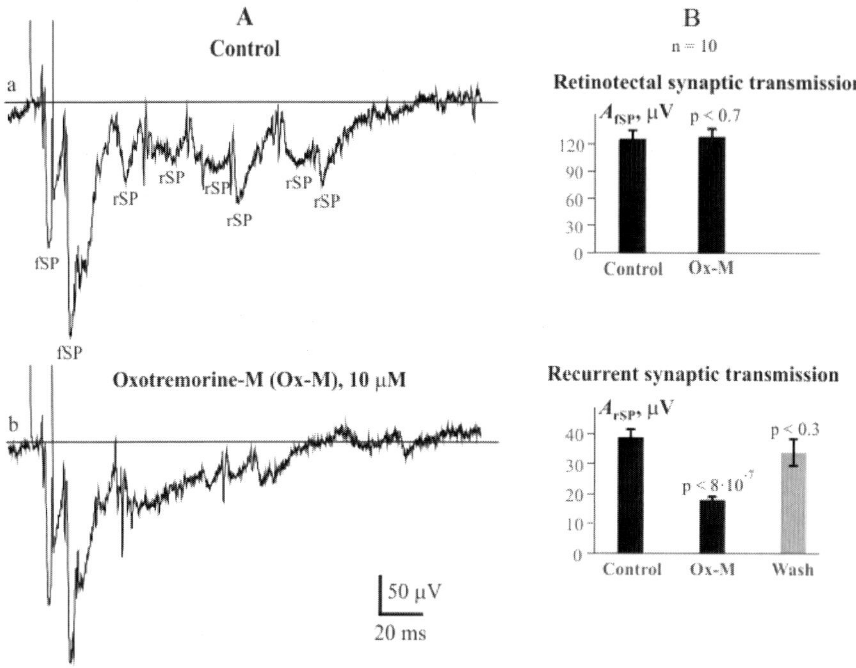

Figure 5. Effect of the specific muscarinic acetylcholine receptor agonist oxotremorine-M (Ox-M) on the retinotectal and recurrent synaptic transmission in the frog tectum. A) The plot shows tectal responses to electrical stimulation of a single retinal ganglion cell or its axon with a two pulses with interpulse interval of 10 ms, recorded in control and Ox-M (10 μM) conditions. The burst of two closely timed action potentials of the ganglion cell elicits in the tectum the fast synaptic potentials (fSPs) and slow negative wave with the recurrent synaptic potentials (rSPs) riding on the back of the wave. The Ox-M (10 μM) has no effect on the fast synaptic potentials (fSPs), generated by the retinotectal synaptic transmission, and greatly depresses the recurrent synaptic potentials (rSPs), generated by the intrinsic recurrent excitatory synaptic transmission. This effect of Ox-M is reversible. B) Average of 10 experiments. A_{fSP} – average amplitude of the fast synaptic potentials, A_{rSP} – average amplitude of the recurrent synaptic potentials.

In the experiments described above, the cholinergic modulation of the synaptic transmission in the frog tectum column has been demonstrated by application of the exogenous agonists – carbamylcholine and oxotremorine-M. Does the endogenous physiological acetylcholine do the same? Results of the experiments described below answer this question.

TONIC NICOTINIC POTENTIATION

A normal brain of animals always contains certain concentration of neuromodulators such as acetylcholine, dopamine, serotonin, noradrenaline and others (Descarries et al., 1997). The frog brain is not exception from this rule. At normal physiological conditions, a certain level of the acetylcholine exists in the frog brain. We have called it the background acetylcholine. Most of the acetylcholine found in an amphibian tectum is supplied by projections from the nucleus isthmi (Desan et al., 1987; Marin and Gonzalez, 1999). The background acetylcholine partly activates acetylcholine receptors of the brain. This phenomenon we have called tonic activation (it can, also, be called steady-state or persistent activation). The tonic activation of the acetylcholine receptors should, in turn, lead to a tonic (steady-state, persistent) modification of synaptic transmission.

We have demonstrated experimentally by application of specific nicotinic acetylcholine receptor antagonist d-tubocurarine (D-TC) (Kuras and Gutmaniene, 2001; Baginskas et al., 2011) that the retinotectal synaptic transmission exhibits the tonic nicotinic potentiation. The idea of the experiment was as follows: If the retinotectal synaptic transmission is persistently enhanced by the action of the background acetylcholine on the presynaptic nAChRs, then the application of nicotinic acetylcholine receptor antagonist D-TC must lead to a decrease of the synaptic potentials and an increase in their paired-pulse facilitation. Results of the experiments are shown in Figure 6.

Perfusion of the tectum surface with 50 μM solution of D-TC led to a considerable decrease of the retinotectal synaptic transmission and to associated increase in the paired-pulse facilitation of the transmission.

Amplitude of the fSPs decreased from the value of 283 ± 26 μV to the value of 155 ± 19 μV, i.e., 1.7 ± 0.1 times.

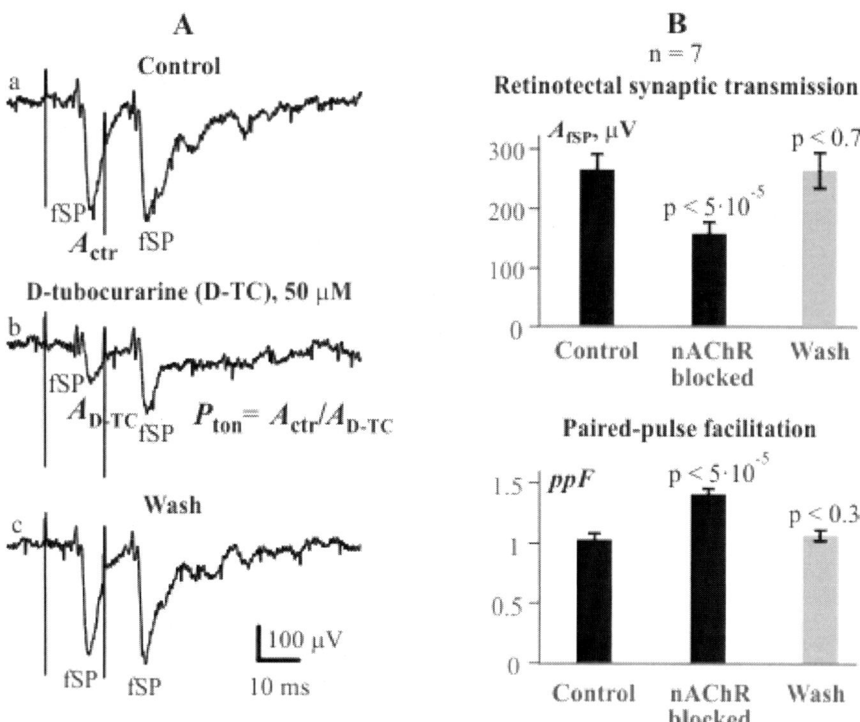

Figure 6. Effect of the specific nicotinic acetylcholine receptor blocker D-tubocurarine (D-TC) on the retinotectal synaptic transmission in the frog tectum. A) The plot shows tectal responses to electrical stimulation of a single retinal ganglion cell or its axon with a pair of pulses with interpulse interval of 15 ms recorded in the control, D-TC (50 μM) and washing conditions. Pair of action potentials of the ganglion cell elicits in the tectum two fast synaptic potentials (fSPs), generated by the retinotectal synaptic transmission. Application of D-TC (50 μM) leads to a substantial decrease of the fSPs. Tonic potentiation, P_{ton}, was estimated as a ratio of amplitudes of the fSPs in the control conditions, A_{ctr}, and in the conditions when nAChRs were blocked by D-TC, $A_{D\text{-}TC}$, i.e., $P_{ton} = A_{ctr}/A_{D\text{-}TC}$. The effect of D-TC was reversible. Paired-pulse facilitation, *ppF*, was estimated as a ratio of amplitudes of the second and first fSPs in the response. B) Average of 7 experiments. A_{fSP} – average amplitude of the fast synaptic potentials, *ppF* – average paired pulse facilitation.

Tonic potentiation, P_{ton}, has been evaluated as a ratio of amplitudes of the fSPs in the control conditions, A_{ctr}, and in conditions, when nicotinic acetylcholine receptors were blocked by application of D-TC, $A_{D\text{-}TC}$, i.e., $P_{ton} = A_{ctr}/A_{D\text{-}TC}$. Concomitantly, the paired-pulse facilitation (*ppF*) of the fSPs increased from the value of 1.03 ± 0.05 to the value of 1.4 ± 0.05, indicating the presynaptic site of action of the D-TC on the fSPs. The effect of D-TC was reversible. Thus, the results of the experiments have demonstrated that the retinotectal synaptic transmission was enhanced 1.7 times due to a persistent activation of the presynaptic nicotinic acetylcholine receptors by the background acetylcholine. This enhancement we have called tonic nicotinic potentiation of the retinotectal synaptic transmission.

Phasic Nicotinic Potentiation

During years of the experiments we have observed that the retinotectal transmission was enhanced for a certain period of time (~1 min) following the sufficiently strong bursts of the action potentials. This burst-induced potentiation we have called phasic potentiation of the retinotectal transmission (the phenomenon can also be called transient after-burst potentiation).

Paradigm of the experiments set to explore this phenomenon is shown if Figure 7. Conditioning stimuli of various strength (2 to 10 pulses with interpulse intervals of 10 ms) have been delivered, and responses recorded. Then, the testing stimuli (2 pulses with interpulse interval of 15 ms) have been delivered at different times – 0.5, 1, 3, 5, 10, 20, 40, 60, 80 s, after the end of conditioning stimulus, and responses recorded. Phasic potentiation, P_{ph}, has been evaluated as a ratio of amplitudes of the testing, A_{test}, and conditioning, A_{cond}, fast synaptic potentials, i.e., $P_{ph} = A_{test}/A_{cond}$.

Results of the experiments are presented in Figure 8. The preceding (5 seconds before) burst of 6 retinotectal action potentials with interpulse intervals of 10 ms caused an increase of the amplitude of the fSP from the value of 167 ± 18 μV to the value of 283 ± 30 μV, i.e., 1.7 ± 0.06 times.

Thus, the phasic potentiation of the retinotectal synaptic transmission was equal to $P_{ph} = A_{test}/A_{cond} = 283/167 = 1.7$. This phasic potentiation of the retinotectal transmission was accompanied by a decrease of paired-pulse facilitation (*ppF*) of the fSPs from the value of 1.35 ± 0.06 to the value of 0.92 ± 0.06, indicating the presynaptic site of generation of the phenomenon and evidencing that the acetylcholine receptors, responsible for the phasic potentiation, are located on presynaptic terminals of the retinotectal fiber.

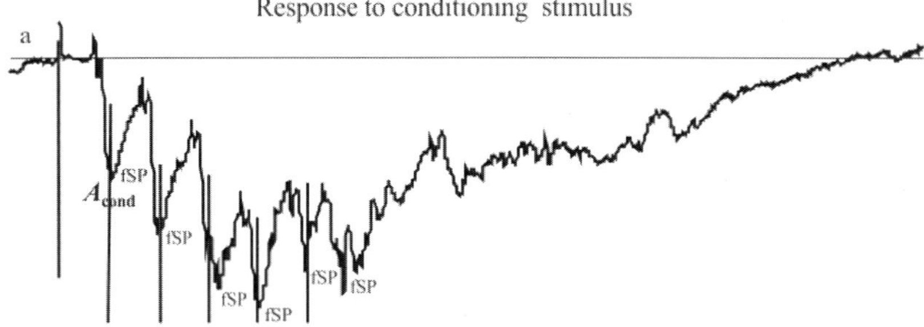

Response to testing stimulus, delivered at a certain time (5 s) after the conditioning burst

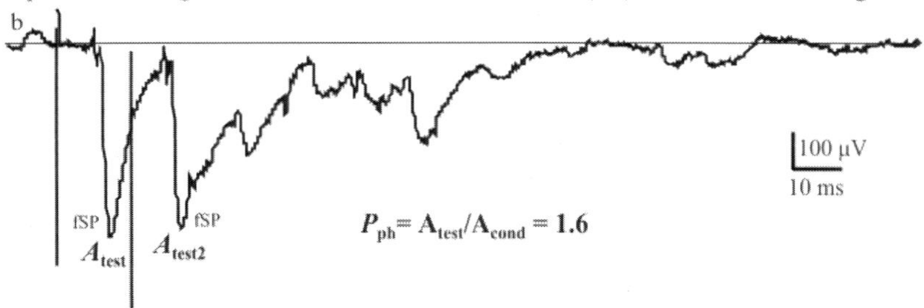

Figure 7. Experimental paradigm for investigation of the phasic potentiation of the retinotectal synaptic transmission. Trace (a) shows tectal response to conditioning stimulation of a single retinal ganglion cell or its axon by a train of 6 pulses at interpulse intervals of 10 ms. Trace (b) shows tectal response to testing stimulation of a single retinal ganglion cell or its axon by a pair of pulses with the interpulse interval of 15 ms. A_{cond} – amplitude of the first conditioning fSP, A_{test} – amplitude of the first testing fSP. The phasic potentiation of the retinotectal synaptic transmission is calculated as a ratio of amplitudes of the testing and conditioning fSPs, $P_{ph} = A_{test}/A_{cond}$.

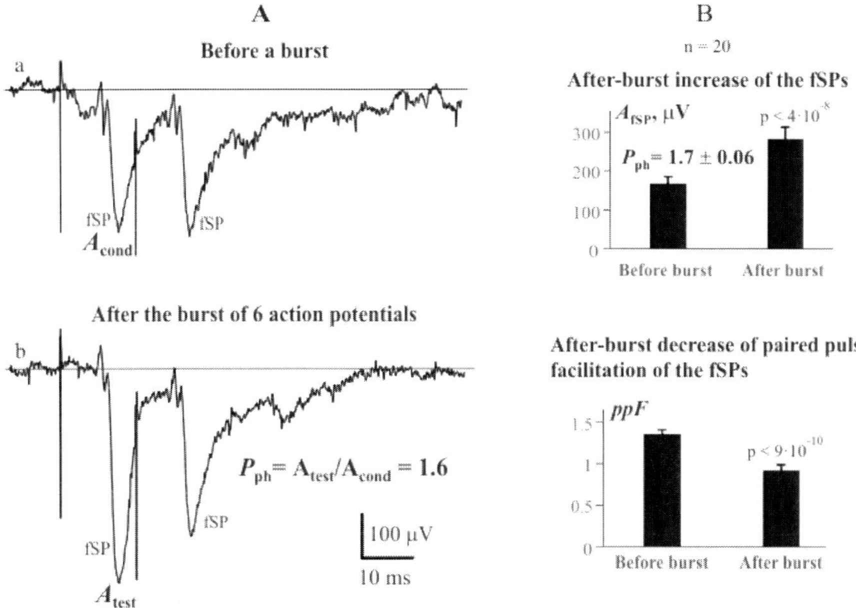

Figure 8. Phasic potentiation of the retinotectal synaptic transmission. A) The plot shows tectal responses to electrical stimulation of a single retinal ganglion cell or its axon with a pair of pulses with interpulse interval of 15 ms, recorded before the conditioning burst of action potentials in the retinotectral fiber (a) and 5 seconds after the burst (b). The conditioning burst consisted of 6 action potentials with interpulse intervals of 10 ms. The phasic potentiation (after-burst increase) of the fast synaptic potentials (fSPs) is evident. The phasic potentiation, P_{ph}, was calculated as a ratio of amplitude of the fSP after the burst, A_{test}, and before the burst, A_{cond}, $P_{ph} = A_{test}/A_{cond}$. Paired-pulse facilitation (*ppF*) of the fSPs was estimated as a ratio of amplitudes of the second and first fSPs of the response. B) Average of 20 experiments. A_{fSP} – average amplitude of the fSPs, *ppF* – average paired pulse facilitation of the fSPs.

Dependence on Strength of the Burst

The phasic potentiation has depended on the burst strength. Results of the experiments are shown in Figure 9. The bursts of 2, 4, 6, 8 and 10 action potentials have been tested for the ability to enhance the retinotectal synaptic transmission. The burst of two action potentials with interpulse interval of 10 ms had no effect on the retinotectal synaptic transmission, $P_{ph} = 1.02 \pm 0.02$.

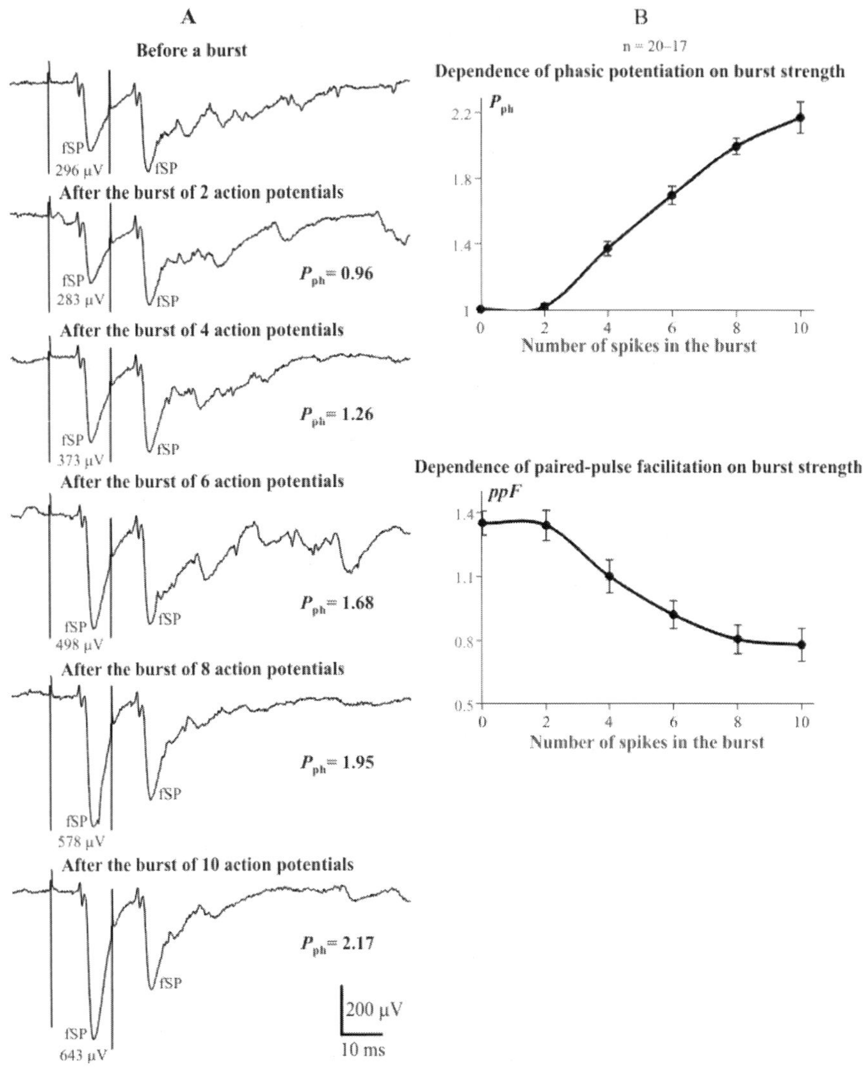

Figure 9. Dependence of the phasic potentiation of the retinotectal transmission on the burst strength. A) The plot shows tectal responses to electrical stimulation of a single retinal ganglion cell or its axon with a pair of pulses with interpulse interval of 15 ms recorded before and 5 seconds after the conditioning bursts of various lengths of retinotectal action potentials. The dependence of the phasic potentiation of the fSPs on the burst strength is evident. The phasic potentiation, P_{ph}, was calculated as a ratio of amplitudes of the fSPs after and before the burst. Paired-pulse facilitation (*ppF*) of the fSPs was estimated as a ratio of amplitudes of the second and first fSPs of the response. B) Average of 20 experiments. P_{ph} – average phasic potentiation of the retinotectal transmission, *ppF* – average paired pulse facilitation of the fSPs.

The bursts of 4, 6, 8 and 10 action potentials have caused a substantial increase of the retinotectal synaptic transmission, P_{ph} = 1.37 ± 0.04, 1.69 ± 0.06, 1.99 ± 0.05, 2.17 ± 0.09, respectively. An increase of the phasic potentiation was accompanied by a decrease of the paired-pulse facilitation (*ppF*) of the retinotectal synaptic transmission, indicating the presynaptic site of generation of the phenomenon. Thus, this is another evidence of localization of the acetylcholine receptors, responsible for the phasic potentiation, on presynaptic terminals of the retinotectal fiber.

Dependence on Time After the Burst

The phasic potentiation depended on the time after the conditioning burst. Time intervals of 0.5, 1, 3, 5, 10, 20, 40, 60 and 80 s have been tested. Results of the experiments are shown in Figure 10.

The values of phasic potentiation of the retinotectal transmission at 0.5, 1, 3, 5, 10, 20, 40, 60 and 80 s after the burst were equal to, P_{ph} = 2.16 ± 0.13, 2.1 ± 0.15, 2.03 ± 0.09, 1.96 ± 0.12, 1.91 ± 0.23, 1.63 ± 0.2, 1.28 ± 0.04, 1.14 ± 0.05, 1.01 ± 0.05, respectively. Thus, the burst induced phasic potentiation has subsided during a period of ~1 min after the burst.

Nicotinic Nature of Phasic Potentiation

What is the kind of acetylcholine receptors, activation of which leads to the phasic potentiation of retinotectal transmission? Experiments with an application of specific nicotinic acetylcholine receptor antagonist d-tubocurarine (D-TC) answer this question. Results of the experiments are shown in Figure 11.

Perfusion of the tectum surface with 50 μM solution of d-tubocurarine led to a decrease of the phasic potentiation of the retinotectal transmission from the value of 1.71 ± 0.09 to the value of 1.12 ± 0.02.

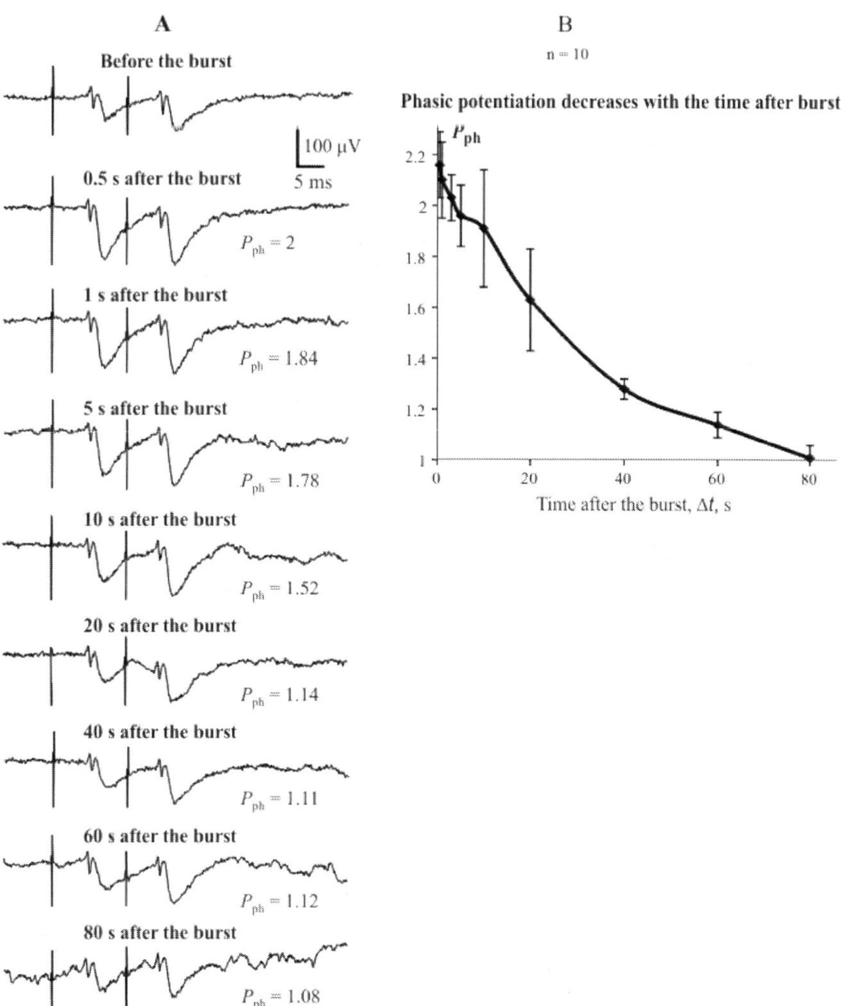

Figure 10. Dependence of phasic potentiation of the retinotectal transmission on the time after the burst. A) The plot shows tectal responses to electrical stimulation of a single retinal ganglion cell or its axon with a pair of pulses with interpulse interval of 15 ms recorded before and at various times after the conditioning burst of action potentials of the retinotectal fiber. The conditioning burst consisted of 8 action potentials with interpulse intervals of 10 ms. The phasic potentiation of the fSPs subsides within the time interval of ~1 min after the burst. B) Average of 10 experiments. P_{ph} – average phasic potentiation of the retinotectal transmission.

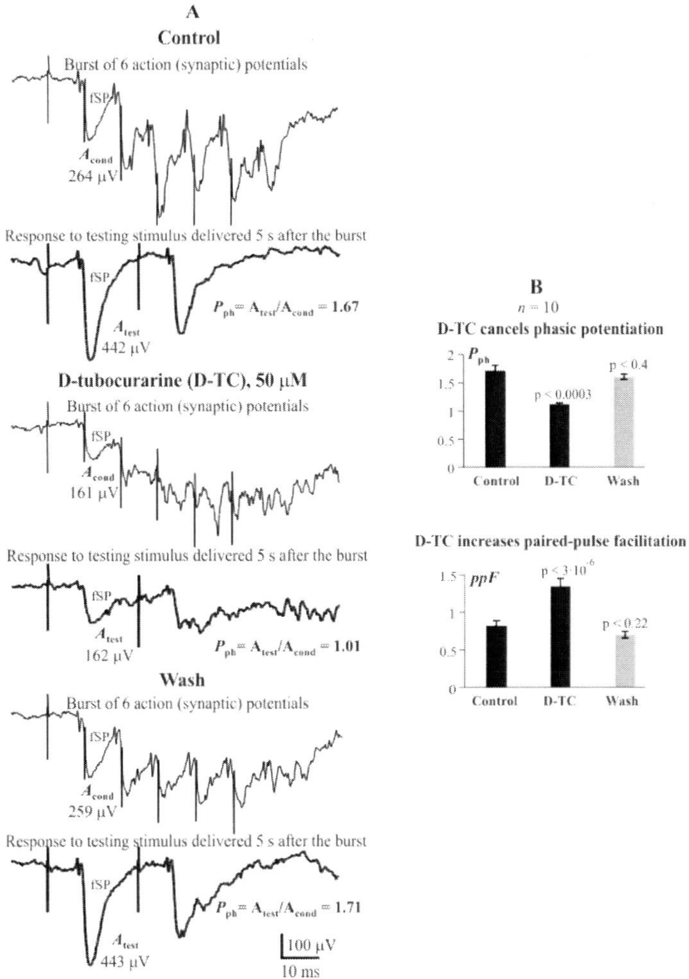

Figure 11. Nicotinic acetylcholine receptor antagonist d-tubocurarine abolishes the phasic potentiation of the retinotectal transmission. A) The plot shows tectal responses to electrical stimulation of a single retinal ganglion cell or its axon with conditioning bursts of 6 pulses with interpulse intervals of 10 ms and testing pairs of pulses with interpulse interval of 15 ms. The responses in the control, D-TC (50 μM) and washing conditions are shown. The phasic potentiation, P_{ph}, was calculated as a ratio of amplitudes of the fast synaptic potentials (fSPs) of the testing and conditioning responses, i.e., $P_{ph} = A_{test}/A_{cond}$, where A_{cond} is amplitude of the first fSPs in the conditioning burst, A_{test} is amplitude of the first fSPs in the testing pair. B) Average of 10 experiments. P_{ph} – average phasic potentiation of the retinotectal transmission, *ppF* – average paired pulse facilitation of the fSPs, calculated as a ratio of amplitudes of the second and first fSPs of the testing pair.

The paired-pulse facilitation of retinotectal transmission has increased from the value of 0.82 ± 0.07 to the value of 1.34 ± 0.11, indicating the presynaptic site of action of D-TC. The effect of D-TC was reversible. Thus, results of the experiments have demonstrated that the phasic potentiation is mediated by the presynaptic nicotinic acetylcholine receptors.

Distinct Receptors Mediate the Phasic and Tonic Potentiation

The presynaptic nicotinic acetylcholine receptors mediate not only the phasic but also the tonic potentiation of the retinotectal transmission (see previous section). Are the presynaptic nicotinic receptors responsible for the phasic potentiation of the same kind as those responsible for the tonic potentiation, or are they different? Results of the experiments described above have suggested different variety of the receptors.

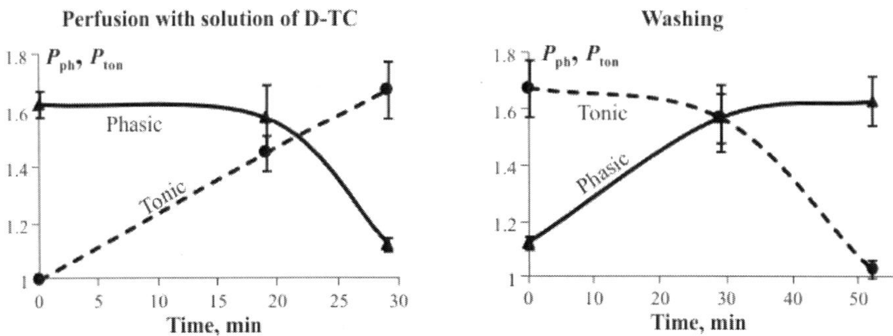

Figure 12. Different time course of the tonic and phasic potentiation during a perfusion with the solution of D-TC and washing. Solid curve – phasic potentiation, P_{ph}, dashed curve – tonic potentiation, P_{ton}.

The tonic and phasic potentiation have been evaluated from the recordings of the responses at various times during a perfusion with the solution of D-TC and washing. The results are shown in Figure 12.

The change (steepness of the slope of the dependence) of the phasic potentiation has lagged behind the change (steepness of the slope of the dependence) of the tonic potentiation during a perfusion with the solution of D-TC, and, contrarily, outran it during washing. This implies that the nicotinic acetylcholine receptors, which account for the tonic potentiation, are of higher affinity to the antagonist d-tubocurarine than the receptors, which account for the phasic potentiation. We have extrapolated this result to the agonist acetylcholine, too, suggesting that the tonic nicotinic acetylcholine receptors are of higher affinity to the acetylcholine than phasic. Such a difference in the features of the tonic and phasic nicotinic receptors conforms well to different functions of the receptors.

CO-MEDIATOR ACETYLCHOLINE

Where the acetylcholine needed for activation of the phasic presynaptic nicotinic receptors comes from? Most logical and relevant answer to this question would be “the presynaptic terminals of the retinotectal fiber”. This answer implies the co-release mechanism. Presynaptic terminals of the retinotectal fiber contain vesicles filled in with the main mediator glutamate and co-mediator acetylcholine. The glutamate and acetylcholine can reside in the same or in separate vesicles. During an excitation of the retinotectal fiber, both glutamate and acetylcholine release from the presynaptic terminals. Glutamate acts as a main mediator. It activates postsynaptic glutamate receptors, generating the retinotectal fast synaptic potentials (fSPs). Acetylcholine acts as a neuromodulator. It activates the presynaptic nicotinic acetylcholine receptors, leading to the phasic potentiation of the retinotectal synaptic transmission (transient after-burst increase of the amplitude of the fSPs). The co-release of glutamate and acetylcholine is not a rare phenomenon in the CNS. It has been found in many of synapses (Li et al., 2004; Nishimaru et al., 2005; Allen et al., 2006; Ren et al., 2011; Hnasko and Edwards, 2012).

However, another possibility that the acetylcholine comes from the cholinergic nuclei of the frog brain cannot be ruled out. For example, it can come from the frog midbrain nucleus isthmi that receives input from the tectum and sends cholinergic axons back to the tectum (Ricciuti and Gruberg, 1985; Titmus et al., 1999). Tectum neurones, which target the nucleus isthmi, can receive input from retinotectal fibers, and can be excited by firing of the retinotectal fibers. This, in turn, leads to an activation of the nucleus isthmi neurones, cholinergic axons of which project back to the tectum. Firing of those axons leads to a release of the acetylcholine into the tectum.

Thus, we have two hypotheses about the source of the acetylcholine that generates the phasic nicotinic potentiation of the retinotectal transmission – co-release hypothesis and nucleus isthmi hypothesis. Below we present evidences corroborating the co-release hypothesis.

First evidence comes from the monotonic character of the dependence of the phasic potentiation on the burst strength (Figure 9) and on the time after the burst (Figure 10). The nucleus isthmi hypothesis to be true the dependence on the burst strength should be threshold-like and/or step-like, and the dependence on the time after the burst should be non-monotonic.

Another evidence corroborating the co-release hypothesis comes from the results of the experiments with an application of glutamate receptor antagonists kynurenic acid and CNQX (Baginskas et al., 2012). The idea behind the experiments was as follows. If the acetylcholine that induces the phasic nicotinic potentiation comes from nucleus isthmi, then a decrease of the synaptic excitation of tectum neurones and consequent decrease of the excitation of nucleus isthmi neurones must lead to a reduction of the phasic potentiation. The opposite result or no effect would prove the co-release hypothesis. Solutions of the kynurenic acid and CNQX of low or moderate concentration have been applied to the tectum to weaken the retinotectal synaptic transmission. Phasic potentiation of the retinotectal transmission has been evaluated and compared to the control value. Results of the experiments are shown in Figures 13, 14.

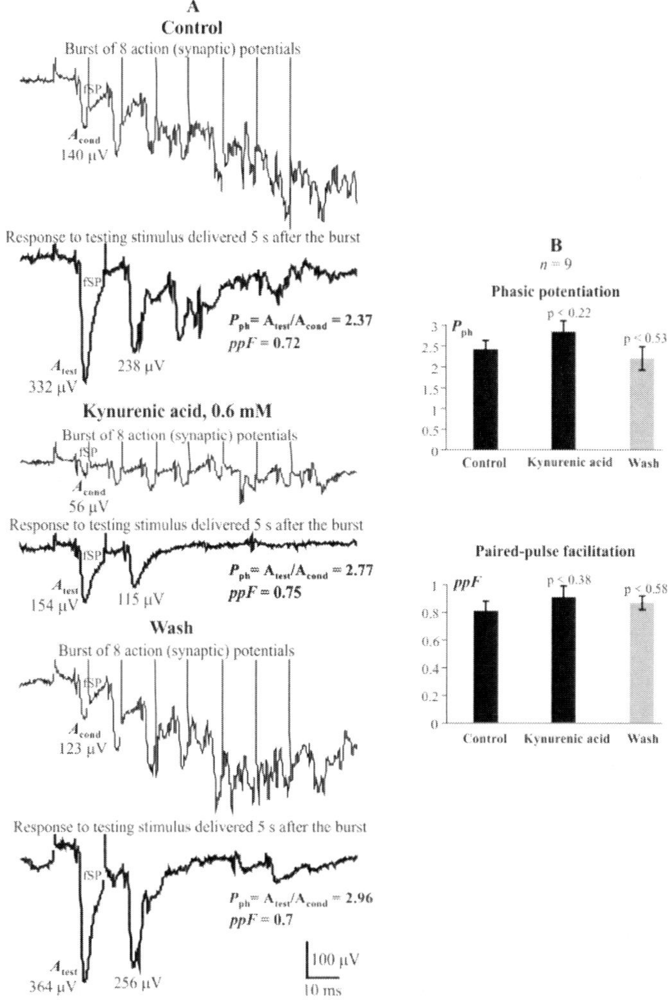

Figure 13. Effect of the application of kynurenic acid on the phasic potentiation of the retinotectal transmission. A) Plot shows tectal responses to electrical stimulation of a single retinal ganglion cell or its axon with the conditioning bursts of 8 pulses with interpulse intervals of 10 ms and testing pairs of pulses with interpulse interval of 15 ms. Responses in the control, kynurenic acid (0.6 mM) and washing conditions are shown. The phasic potentiation, P_{ph}, was calculated as a ratio of amplitudes of the fast synaptic potentials (fSPs) of the testing and conditioning responses, i.e., $P_{ph} = A_{test}/A_{cond}$,where A_{cond} is amplitude of the first fSP in the conditioning burst, A_{test} is amplitude of the first fSP in the testing pair. B) Average of 9 experiments. P_{ph} – averge phasic potentiation of the retinotectal transmission, *ppF* – average paired pulse facilitation of the fSPs, calculated as a ratio of amplitudes of the second and first fSPs in the testing pair.

Perfusion of the tectum surface with a low (0.6 mM) concentration solution of the kynurenic acid led to a decrease of the retinotectal synaptic potentials from the value of 155 ± 21 μV to the value of 43 ± 6 μV, i.e., 3.6 times. The phasic potentiation (P_{ph}) of the retinotectal synaptic transmission in the control and kynurenic acid conditions was equal to 2.42 ± 0.21 and 2.84 ± 0.26, respectively. Hence, in the conditions of reduced synaptic excitation (through the application of kynurenic acid), the phasic potentiation has slightly (although insignificantly) increased. The paired-pulse facilitation (*ppF*) in the control and kynurenic acid conditions was equal to 0.81 ± 0.07 and 0.91 ± 0.08, respectively, i.e., have not differed significantly.

Perfusion of the tectum surface with a low (5 – 10 μM) concentration solution of the CNQX led to a decrease of the retinotectal synaptic potentials from the value of 131 ± 14 μV to the value of 23 ± 5 μV, i.e., 5.7 times. The phasic potentiation of the retinotectal synaptic transmission in the control and CNQX conditions was equal to 2.31 ± 0.21 and 2.28 ± 0.18, respectively. The paired-pulse facilitation in the control and CNQX conditions was equal to 0.88 ± 0.04 and 0.99 ± 0.06, respectively. Hence, in the conditions of reduced synaptic excitation (through the application of CNQX), the phasic potentiation and paired-pulse facilitation of the retinotectal synaptic transmission have not differed significantly from the values in normal conditions.

Thus, the results of the experiments have clearly demonstrated that despite the almost six-fold decrease of the excitation of the tectum neurons, the phasic potentiation and paired pulse facilitation of the retinotectal synaptic transmission remained the same, ruling out the nucleus isthmi hypothesis and corroborating the co-release hypothesis.

During firing of the retinotectal fiber, the acetylcholine releases as a co-mediator from the presynaptic terminals (***a*** and ***b*** synapses in Figure 1) together with the main mediator glutamate. The co-released acetylcholine acts as a neuromodulator. It activates the presynaptic nicotinic acetylcholine receptors located in the terminals of retinotectal fiber, leading to the phasic potentiation of the retinotectal synaptic transmission, or, saying in other words, to a transient after-burst increase of the release of the main mediator glutamate.

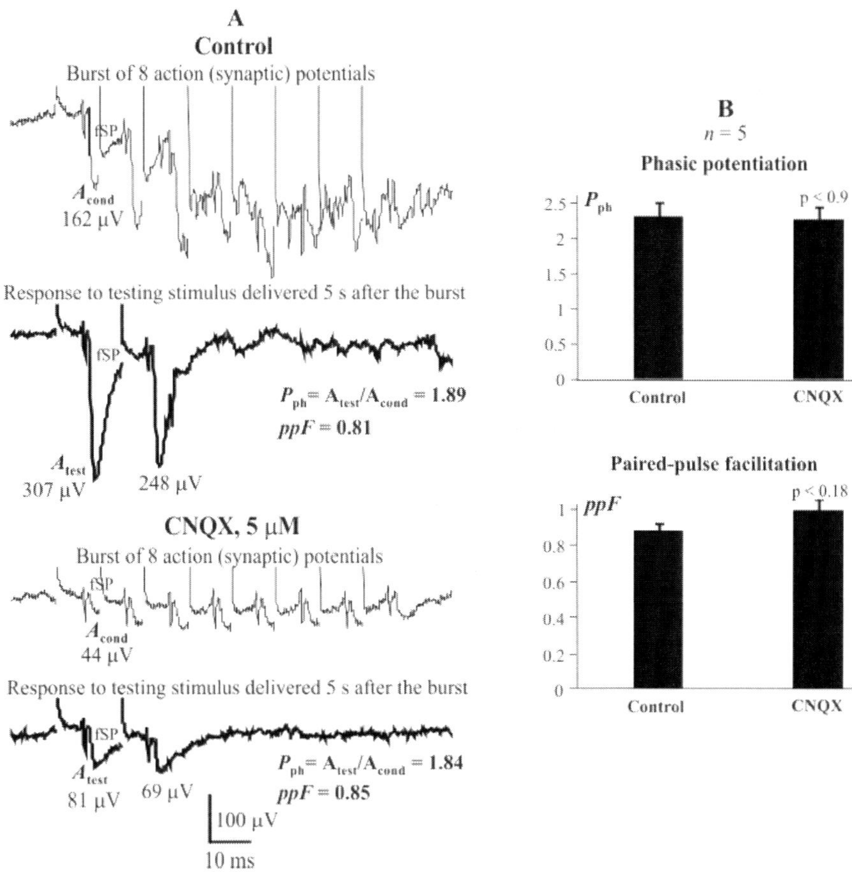

Figure 14. Effect of the application of CNQX on the phasic potentiation of the retinotectal transmission. A) Plot shows tectal responses to electrical stimulation of a single retinal ganglion cell or its axon with the conditioning bursts of 8 pulses with interpulse intervals of 10 ms and testing pairs of pulses with interpulse interval of 15 ms. Responses in the control and CNQX (5 µM) conditions are shown. The phasic potentiation, P_{ph}, was calculated as a ratio of amplitudes of the fast synaptic potentials (fSPs) of the testing and conditioning responses, i.e., $P_{ph} = A_{test}/A_{cond}$, where A_{cond} is amplitude of the first fSP in the conditioning burst, A_{test} is amplitude of the first fSP in the testing pair. B) Average of five experiments. P_{ph} – average phasic potentiation of the retinotectal transmission, *ppF* – average paired pulse facilitation of the fSPs, calculated as a ratio of amplitudes of the second and first fSPs in the testing pair.

Muscarinic Inhibition of Recurrent Excitation

In the study (Baginskas and Kuras, 2016), we have demonstrated that the co-mediator acetylcholine modulates not only the retinotectal synaptic transmission (***a*** and ***b*** synapses in Figure 1) but also the tectum intrinsic recurrent excitatory synaptic transmission (***d*** and ***c*** synapses in Figure 1). However, this modulation has had an opposite sign and occurred with a delay.

The experiments revealed that recurrent synaptic potentials (rSPs) riding on the earlier part of the slow negative wave (sNW) were always larger than the ones riding on the later part of the sNW. The time moment that separated the earlier and later parts of the sNW coincided with the time moment of occurrence of absolute maximal value of the slow negative potential (sNP) (usually ~80 ms from the beginning of the response). Experiments were done as follows. Strong stimulus consisting of 5–8 pulses with interpulse intervals of 10 ms has been delivered to evoke the sNP. Time moment of occurrence of the minimum (maximal absolute value) of the sNP, $t_{\mathrm{min,sNP}}$, has been determined. Average value of $t_{\mathrm{min,sNP}}$, evaluated from 25 experiments, was equal to 79 ± 1 ms. Then, weak or moderate stimulus consisting of 2–4 pulses with interpulse intervals of 10–15 ms has been applied to evoke the sNW with the rSPs superimposed on the sNW. Amplitudes of the rSPs have been measured. Average amplitude of the rSPs riding the early part of the sNW (i. e., occurring before the time moment $t_{\mathrm{min,sNP}}$), $\bar{A}_{\mathrm{rSP,early}}$, and average amplitude of the rSPs riding the late part of the sNW (i.e., occurring after the time moment $t_{\mathrm{min,sNP}}$), $\bar{A}_{\mathrm{rSP,late}}$, have been calculated. Delayed inhibition of the recurrent excitation, I_{del}, has been evaluated as a ratio of the average amplitudes of the early and late rSPs, $I_{\mathrm{del}} = \bar{A}_{\mathrm{rSP,early}}/\bar{A}_{\mathrm{rSP,late}}$. Results of the experiments are presented in Figure 15.

Average amplitude of the early rSPs was equal to 62.1 ± 5.4 μV, and average amplitude of the late rSPs was equal to 28.3 ± 2.2 μV. Thus, the delayed inhibition of the recurrent excitation was of 2.21 ± 0.07 times. The delay of the inhibition was estimated to be equal to 79 ± 1 ms.

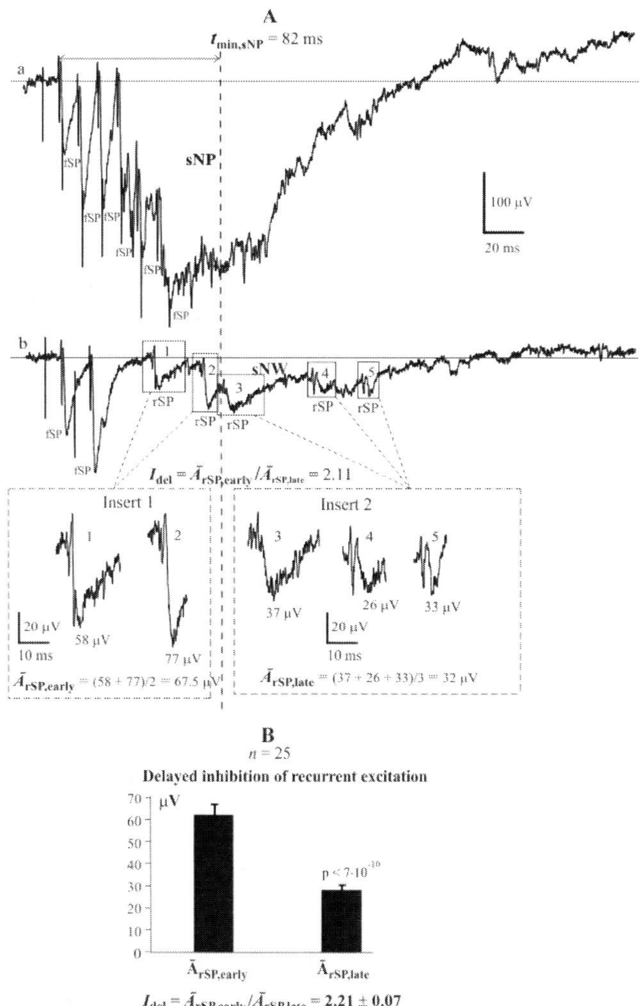

Figure 15. Delayed inhibition of the recurrent excitation. A) Plot shows tectal responses to electrical stimulation of a single retinal ganglion cell or its axon with the burst of 6 pulses with interpulse intervals of 10 ms and with two pulses with interpulse interval of 15 ms. The strong stimulation elicits in the tectum the fast synaptic potentials (fSPs) and slow negative potential (sNP). The weaker stimulation elicits the fSPs and slow negative wave (sNW) with the recurrent excitatory synaptic potentials (rSPs) superimposed on the sNW. The time moment of occurrence of the minimum (maximal absolute value) of the sNP, $t_{min,sNP}$, separates the early and late parts of the sNW. Insert 1 shows the rSPs on the early part of the sNW. Insert 2 shows the rSPs on the late part of the sNW. $\bar{A}_{rSP,early}$ – average amplitude of the early rSPs. $\bar{A}_{rSP,late}$ – average amplitude of the late rSPs. I_{del} – delayed inhibition of the recurrent excitation. B) Average of 25 experiments.

Muscarinic Nature of Delayed Inhibition

What is the source of the delayed inhibition of the recurrent excitation? Results of the experiments with an application of muscarinic acetylcholine receptor antagonist atropine have demonstrated that the delayed inhibition of the recurrent excitation is mediated by the activation of muscarinic acetylcholine receptors. Effect of the atropine on the amplitudes of early and late rSPs have been measured. The results are shown in Figure 16.

Perfusion of the tectum surface with 50 μM concentration solution of the atropine had no significant effect on the average amplitude of the early rSPs, ($\bar{A}_{rSP,early}$ = 59.9 ± 11.6 μV (control), 50.1 ± 4.1 μV (atropine)), and substantially increased the average amplitude of the late rSPs from the value of 27.2 ± 4.5 μV to the value of 39.9 ± 3.9 μV. Consequently, the atropine has substantially decreased the delayed inhibition of the recurrent excitation from the value of 2.16 ± 0.07 to the value of 1.27 ± 0.05. Thus, the results of the experiments have demonstrated that the delayed inhibition of the recurrent excitation is mediated by the muscarinic acetylcholine receptors. They can be located on the presynaptic terminals of recurrent pear-shaped neurons (***c***, ***d*** synapses in Figure 1) and/or in the membrane of postsynaptic neuron (**PC** neuron in Figure 1).

The acetylcholine that activates the muscarinic receptors releases, presumably, through the same mechanism as the acetylcholine that activates the phasic presynaptic nicotinic receptors responsible for the phasic potentiation of the retinotectal transmission, i.e., through the co-release mechanism, addressed in the preceding section. However, the nicotinic receptors were activated without any delay, muscarinic – with ~80 ms delay. The time needed for diffusion of the acetylcholine towards the muscarinic receptors cannot account for the delay of muscarinic inhibition, since the retinotectal and recurrent fiber synapses are located in the same region (see Figure 1). So, how can the delay be explained?

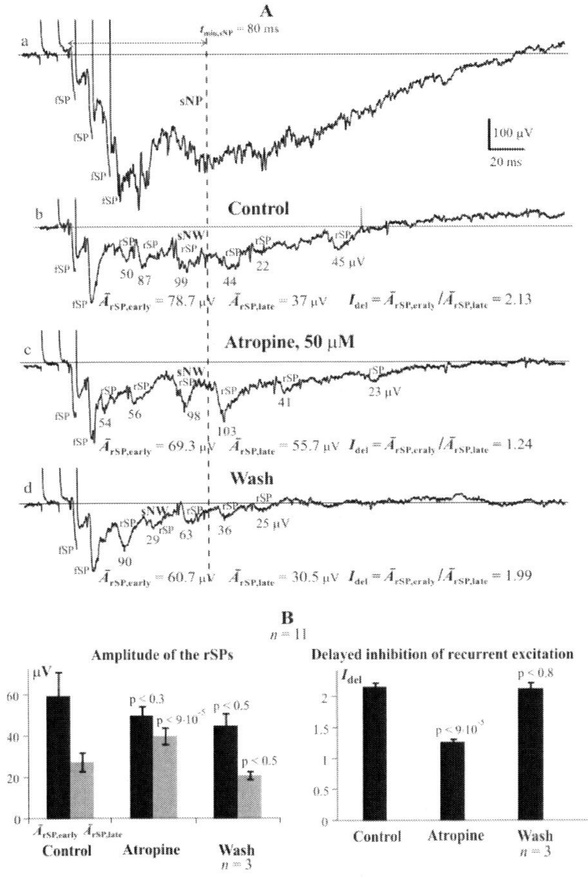

Figure 16. Effect of muscarinic acetylcholine receptor blocker atropine on the delayed inhibition of the recurrent excitation. A) Plot shows tectal responses to bursts of four (a) and two (b, c, d) action potentials of the retinotectal fiber with the interpulse intervals of 10 ms. The burst of four action potentials elicits fast synaptic potentials (fSPs) and slow negative potential (sNP). The burst of two action potentials elicits the fSPs and slow negative wave (sNW) with the recurrent excitatory synaptic potentials (rSPs) superimposed on the sNW. Recordings in the control (b), atropine (50 µM) (c) and washing (d) conditions are shown. The time moment of occurrence of the minimum (maximal absolute value) of the sNP, $t_{min,sNP}$, separates the early and late parts of the sNW. Amplitudes of the rSPs were measured, and average amplitudes of the rSPs on the early and late parts of the sNW, $\bar{A}_{rSP,early}$, $\bar{A}_{rSP,late}$, respectively, were calculated. The delayed inhibition of the recurrent excitation, I_{del}, was evaluated as a ratio of these amplitudes, $I_{del} = \bar{A}_{rSP,early}/\bar{A}_{rSP,late}$. B) Average of 11 experiments. Black colored columns show average amplitude of the early rSPs, gray colored columns show average amplitude of the late rSPs.

Explanation comes from the results of biochemical and electrophysiological studies demonstrating that activation of the nicotinic receptors leads to an effect immediately after attaching of the agonist to the receptor, the activation of muscarinic receptors leads to an effect with a 50–100 ms delay after attaching of the agonist. The activation of muscarinic acetylcholine receptors does not lead to an effect directly because intermediate signal transduction stages through G-proteins and secondary messenger systems are involved in the process (Brown, 2010). Completion of the transduction steps takes a time that can account for 80 ms delay. For example, the FRET sensor method of monitoring of G protein-coupled receptor activation estimates 80–100 ms delay of the effect mediated by the activation of muscarinic receptors of subtypes M1, M2 and M3 (Falkenburger et al., 2010; Hoffmann et al., 2012; Maier-Peuschel et al., 2010), which is very close to our estimation. Electrophysiological recordings from sympathetic neurons in the bullfrog have demonstrated that muscarinic inhibition follows cholinergic excitation with a delay of 50–100 ms (Dodd and Horn, 1983). Hasselmo and Fehlau (2001) have reported that activation of muscarinic acetylcholine receptors in the rat hippocampus causes presynaptic inhibition of glutamatergic synaptic potentials at excitatory feedback connections with the onset time constant between 1 and 2 s. Therefore, other investigators with different methods and models got similar values for the delay of muscarinic inhibition as it was found in our experiments.

CONCLUSION

Results of the experiments described above have demonstrated the neuromodulating role of the acetylcholine in the frog tectum. The acetylcholine enhances retinotectal excitatory synaptic transmission, and depresses intrinsic recurrent excitatory synaptic transmission.

A certain level of ambient acetylcholine (background acetylcholine) exists in the tectum due to continuous activity of cholinergic nuclei in the frog brain, primarily the nucleus isthmi. This background acetylcholine causes the activation of tonic presynaptic nicotinic receptors, leading to a tonic potentiation of the retinotectal (optic) synaptic transmission up to 1.7 times.

The main mediator glutamate and co-mediator acetylcholine release into the extracellular space during firing of the retinotectal (optic) fiber. The co-released acetylcholine acts as a neuromodulator. Specifically, it activates presynaptic phasic nicotinic receptors, leading to a phasic (after-burst) nicotinic potentiation of the retinotectal transmission up to 2.2 times. Magnitude of the phasic nicotinic potentiation depends on the intensity of the burst of action potentials and on the time after the burst. It subsides within a period of ~1 min after the burst.

Distinct presynaptic nicotinic receptors mediate the tonic and phasic potentiation of the retinotectal transmission differing in their affinity to the antagonist d-tubocurarine. The tonic nicotinic receptors are of higher affinity to the d-tubocurarine than the phasic receptors. This result we have extrapolated to the agonist acetylcholine, too, suggesting that the tonic nicotinic receptors are of higher affinity to the acetylcholine than the phasic ones. This conforms well to different functions of the receptors.

The retinotectal synaptic excitation activates recurrent excitatory neural circuits in the tectum, which constitute a positive feedback in the tectum column. The co-mediator acetylcholine, released from terminals of the retinotectal fiber during its firing, modulates excitatory synaptic transmission in these recurrent circuits. However, this modulation is of opposite sign (depression) than the modulation of the retinotectal synaptic transmission (potentiation). The acetylcholine co-released during a burst of the retinotectal fiber activates muscarinic receptors located in the presynaptic terminals of recurrent pear-shaped neurones and/or in the membrane of postsynaptic neurones, leading to a depression of the recurrent excitatory potentials up to 2 times. This muscarinic inhibition of the recurrent excitation occurs with a ~80 ms delay, in agreement with the results of biochemical studies showing that the G-protein coupled effectors exert their effects with 80–100 ms delay.

References

Allen, T. G., Abogadie, F. C., Brown, D. A. 2006. Simultaneous release of glutamate and acetylcholine from single magnocellular "cholinergic" basal forebrain neurons. *J. Neurosci.*, 26: 1588-1595.

Baginskas, A., Kuraite, V., Kuras, A. 2011. Presynaptic nicotinic potentiation of a frog retinotectal transmission evoked by discharge of a single retina ganglion cell. *Neurosci. Res.*, 70: 391-400.

Baginskas, A., Kuraite, V., Kuras, A. 2012. Frog retinal ganglion cells projecting to the tectum layer F release acetylcholine as co-mediator. *Neurosci. Lett.*, 522: 145-150.

Baginskas, A., Kuras, A. 2008. Single retina ganglion cell evokes the activation of L-type Ca^{2+}-mediated slow inward current in frog tectal pear-shaped neurons. *Neurosci. Res.*, 60: 412-421.

Baginskas, A., Kuras, A. 2009. L-Type Ca2+ current in frog tectal recurrent neurons determines the NMDA receptor activation on efferent neuron. *Exp. Brain Res.*, 193: 509-517.

Baginskas, A., Kuras, A. 2011. Muscarinic inhibition of recurrent glutamatergic excitation in frog tectum column prevents NMDA receptor activation on efferent neuron. *Exp. Brain Res.*, 208: 323-334.

Baginskas, A., Kuras, A. 2014. Mechanisms of suprathreshold excitation of a frog tectal neuron column by discharge of a single moving edge or darkness detector and their relation to a frog escape reactions. In *Frogs Genetic Diversity, Neural Development and Ecological Implications*, edited by Henry Lambert, 89-137. New York: Nova Science Publishers.

Baginskas, A., Kuras, A. 2016. Retinal co-mediator acetylcholine evokes muscarinic inhibition of recurrent excitation in frog tectum column. *Neurosci. Lett.*, 629: 137-142.

Brown, D. A. 2010. Muscarinic acetylcholine receptors (mAChRs) in the nervous system: some functions and mechanisms. *J. Mol. Neurosci.*, 41(3): 340-346.

Chung, S. H., Bliss, T. V. P., Keating, M. J. 1974. The synaptic organization of optic afferents in the amphibian tectum. *Proc. Roy. Soc. B* (London), 187: 421-447.

Desan, P. H., Gruberg, E. R., Grewell, K. M., Eckenstein, F. 1987. Cholinergic innervations of the optic tectum in the frog *Rana pipiens. Brain Res.*, 413: 344-349.

Descarries, L., Gisiger, V., Steriade M. 1997. Diffuse transmission by acetylcholine in the CNS. *Prog. Neurobiol.*, 53(5): 603-625.

Dodd, J., Horn, J. P. 1983. Muscarinic inhibition of sympathetic C neurones in the bullfrog. *J. Physiol.*, 334: 271-291.

Falkenburger, B. H., Jensen, J. B., Hille, B. 2010. Kinetics of M1 muscarinic receptor and G protein signaling to phospholipase C in living cells. *J. Gen. Physiol.*, 135(2): 81-97.

Forslid, A., Ingvar, M., Rosen, I., Ingvar, D. H. 1986. Carbon dioxide narcosis: influence of short-term high concentration carbon dioxide inhalation on EEG and cortical evoked responses in the rat. *Acta Physiol. Scand.*, 127: 281-287.

George, S. A., Marks, W. B. 1974. Optic nerve terminal arborizations in the frog: shape and orientation inferred from electrophysiological measurements. *Exp. Neurol.*, 42: 467-482.

Gutman, A. M., Kuras, A. V. 1974. Studies on extracellular PSP of single afferent - the quantum of the EEG in tectum opticum of the frog (in Russian). *Biofizika*, 19: 894-898.

Hasselmo, M. E., Fehlau, B. P. 2001. Differences in time course of ACh and GABA modulation of excitatory synaptic potentials in slices of rat hippocampus. *J. Neurophysiol.*, 86(4): 1792-1802.

Hnasko, T. S., Edwards, R. H. 2012. Neurotransmitter corelease: mechanism and physiological role. *Ann. Rev. Physiol.*, 74: 225-243.

Hoffmann, C., Nuber, S., Zabel, U., Ziegler, N., Winkler, C., Hein, P., Berlot, C. H., Bünemann, M., Lohse, M. J. 2012. Comparison of the activation kinetics of the M3 acetylcholine receptor and a constitutively active mutant receptor in living cells. *Mol. Pharmacol.*, 82(2): 236-245.

Jia, F., Chandra, D., Homanics, G. E., Harrison, N. L. 2008. Ethanol modulates synaptic and extrasynaptic GABAA receptors in the thalamus. *J. Pharmacol. Exp. Ther.*, 326: 475-482.

Kealy, J., Commins, S. 2010. Frequency-dependent changes in synaptic plasticity and brain-derived neurotrophic factor (BDNF) expression in the CA1 to perirhinal cortex projection. *Brain Res.*, 1326: 51-61.

King, J. G. Jr., Lettvin, J. Y., Gruberg, E. D. 1999. Selective, unilateral, reversible loss of behavioral responses to looming stimuli after injection of tetrodotoxin or cadmium chloride into the frog optic nerve. *Brain Res.*, 841: 20-26.

Kuras, A. V., Khusainoviene, N. P. 1986. Paired-pulse facilitation of mass extracellular synaptic potentials of separate retinal afferents in the frog tectum (in Russian). *Neirofiziologija*, 18: 45-55.

Kuras, A., Baginskas, A., Batuleviciene, V. 2004. Suprathreshold excitation of frog tectal neurons by short spike trains of single retinal ganglion cell. *Exp. Brain Res.*, 159: 509-518.

Kuras, A., Baginskas, A., Batuleviciene, V. 2006. Non-NMDA and NMDA receptors are involved in suprathreshold excitation of network of frog tectal neurons by a single retinal ganglion cell. *Neurosci. Res.*, 54: 328-337.

Kuras, A., Baginskas, A., Batuleviciene, V., Lamanauskas, N. 2007. Single retinal changing contrast (3rd) detector elicits NMDA receptor response and higher activity level of frog tectum neuron network. *Exp. Brain Res.*, 179: 209-217.

Kuras, A., Gutmaniene, N. 1995. Preparation of carbon-fibre microelectrode for extracellular recording of synaptic potentials. *J. Neurosci. Methods*, 62: 207-212.

Kuras, A., Gutmaniene, N. 1997. Multi-channel metallic electrode for threshold stimulation of frog's retina. *J. Neurosci. Methods*, 75: 99-102.

Kuras, A., Gutmaniene, N. 2001. N-cholinergic facilitation of glutamate release from an individual retinotectal fiber in frog. *Vis. Neurosci.*, 18: 549-558.

Lara, R., Arbib, M. A., Cromarty, A. S. 1982. The role of the tectal column in facilitation of amphibian prey-catching behavior: a neural model. *J. Neurosci.*, 2: 521-530.

Li, W. C., Soffe, S. R., Roberts, A. 2004. Glutamate and acetylcholine corelease at developing synapses. *Proc. Nat. Acad. Scien. USA*, 101: 15488-15493.

Luscher, H-R., Ruenzel, P. W., Henneman, E. (1983). Effects of impulse frequency, PTP, and temperature on responses elicited in large populations of motoneurons by impulse in single Ia-fibres. *J. Neurophysiol.*, 50: 1045-1058.

Maier-Peuschel, M., Frölich, N., Dees, C., Hommers, L. G., Hoffmann, C., Nikolaev, V. O., Lohse, M. J. 2010. A fluorescence resonance energy transfer-based M2 muscarinic receptor sensor reveals rapid kinetics of allosteric modulation. *J. Biol. Chem.*, 285(12): 8793-8800.

Marin, O., Gonzalez, A. 1999. Origin of tectal cholinergic projections in amphibians: a combined study of choline acetyltransferase immunohistochemistry and retrograde transport of dextran amines. *Vis. Neurosci.*, 16: 271-283.

Maturana, H. R., Lettvin, J. Y., McCulloch, W. S., Pitts, W. H. 1960. Anatomy and physiology of vision in the frog (*Rana pipiens). J. Gen. Physiol.*, 43: 129-175.

Nishimaru, H., Restrepo, C. E., Ryge, J., Yanagawa Y.,, Kiehn, O. 2005. Mammalian motorneurons corelease glutamate and acetylcholine at central synapses. *Neurosci.*, 102: 5245-5249.

Obermayer, J., Verhoog, M. B., Luchicchi, A., Mansvelder, H. D. 2017. Cholinergic Modulation of Cortical Microcircuits Is Layer-Specific: Evidence from Rodent, Monkey and Human Brain. *Front. Neural Circuits*, 11: 100.

Parnas, H., Segel, L. A. 1982. Ways to discern the presynaptic effect of drugs on neurotransmitter release. *J. Theor. Biol.*, 94(4): 923-941.

Picciotto, M. R., Higley, M. J., Mineur, Y. S. 2012. Acetylcholine as a neuromodulator: cholinergic signaling shapes nervous system function and behavior. *Neuron*, 76(1): 116-129.

Raastad, M. 1995. Extracellular activation of unitary excitatory synapses between hippocampal CA3 and CA1 pyramidal cells. *Eur. J. Neurosci.*, 7: 1882-1888.

Ren, J., Qin, C., Hu, F., Tan, J., Qiu, L., Zhao, S., Feng, G., Luo, M. 2011. Habenula "cholinergic" neurons co-release glutamate and acetylcholine and activate postsynaptic neurons via distinct transmission modes. *Neuron*, 69: 445-452.

Ricciuti, A. J., Gruberg, E. R. 1985. Nucleus isthmi provides most tectal choline acetyltransferase in the frog *Rana pipiens*. *Brain Res.*, 341: 399-402.

Stern, P., Edwards, F. A., Sakmann, B. 1992. Fast and slow components of unitary EPSCs on stellate cells elicited by focal stimulation in slices of rat visual cortex. *J. Physiol.*, 449: 247-278.

Szekely, G., Lazar, G. 1976. Cellular and synaptic architecture of the optic tectum. In: *Frog Neurobiology*, edited by R. Llinas and W. Precht, 406-433. Berlin, Heidelberg, New York: Springer-Verlag.

Titmus, M. J., Tsai, H. J., Lima, R., Udin, S. B. 1999. Effects of choline and other nicotinic agonists on the tectum of juvenile and adult Xenopus frogs: a patch-clamp study. *Neurosci.*, 91: 753-769.

Walz, C., Elssner-Beyer, B., Schubert, D., Gottmann, K. 2010. Properties of glutamatergic synapses in immature layer Vb pyramidal neurons: coupling of pre-and postsynaptic maturational states. *Exp. Brain Res.*, 200: 169-182.

Yamamoto, K., Koyanagi, Y., Koshikawa, N., Kobayashi, M. 2010. Postsynaptic cell type-dependent cholinergic regulation of GABAergic synaptic transmission in rat insular cortex. *J. Neurophysiol.*, 104: 1933-1945.

In: Acetylcholine Receptors …
Editor: Adelais Eros Gupta

ISBN: 978-1-53615-447-4

Chapter 2

M_4 MUSCARINIC RECEPTORS – STRUCTURE, LIGANDS, DETECTION AND FUNCTION

Jaromir Myslivecek*
Institute of Physiology, 1st Faculty of Medicine,
Charles University, Prague, Czech Republic

ABSTRACT

Muscarinic receptors belong to the class of G protein-coupled receptors and exist in five subtypes (M_1-M_5). M_4 muscarinic receptors (M_4 MR) are those that are together with M_2 MR coupled to G_i protein and thus inhibit adenylyl cyclase. Even-numbered muscarinic receptors are considered primarily as pre-synaptic receptors. However, both M_2 MR and M_4 MR are localized both pre-synaptically and post-synaptically. As cholinergic autoreceptors, M_2 and M_4 provide feedback control of acetylcholine release. M_4 MR are coupled to $G_{i/o}$ G proteins and to G_s, which provides further level of signal fine tuning. M_4 MR are spontaneously active and can cause constitutive inhibition of adenylyl cyclase. M_4 MR were also shown to enhance Ca^{2+} currents via the effect on Ca_{V1} channels. Sequencing study on rat M_4 MR has shown the presence of cell type-specific silencer element in the promoter region and

* Corresponding Author's E-mail: jmys@lf1.cuni.cz.

that expression of M_4 MR is regulated by the neuron-restrictive silencer element/repressor element 1. All MR reveal relatively high amino acid sequence similarity and thus only a few MR ligands are selective. This makes the detection of MR in respective tissue sometimes problematic. Typically, it is not possible to use specific radioligands. Detection of protein – using western blot or immunohistochemical methods – suffers from unexpectedly high non-specificity of antibodies as verified on knockout mice. Similarly, the evaluation of gene expression is also not an appropriate method for M_4 MR detection as the mRNA level usually does not correspond with receptor levels. M_4 MR have been associated with various organism functions during the past years. Initially, the role of MR could be elucidated only by means of pharmacological studies. These studies have indicated the important role of MR in several brain processes including learning and memory, attention, locomotion, thermoregulation, sleep and wakefulness, food intake and reward. Here, I will review the function of M_4 MR. Next, possible detection methods (radioligand binding, autoradiography, western blotting, gene expression) will be discussed with special interest to knockout mice as specific M_4 MR function model. Next, M_4 MR regulated locomotion will be reviewed together with the very recent data on consequences to biorhythms. To sum up, it is important to remember that M_4 MR are a subtype of MR with various functions and connection to biorhythms.

Keywords: muscarinic receptors, central nervous system, biorhythm, motor activity

1. Introduction

Muscarinic receptors belong to the class of G protein-coupled receptors and exist in five subtypes (M_1-M_5) (Thal et al. 2016, Brown 2010). This class of receptors can be further divided into odd-numbered (M_1, M_3, M_5) and even-numbered (M_2 and M_4) subtypes, that activate similar intracellular signaling pathways (Brown 2010). While odd-numbered muscarinic receptors are mainly coupled to G_q protein and thus activate phospholipase C – inositoltrisphosphate – diacylglycerol pathway(s), even-numbered muscarinic receptors are primarily coupled to G_i protein and this inhibits adenylyl cyclase (Myslivecek, Farar, and Valuskova 2017). Not only these main G proteins are activated after specific muscarinic receptor subtype activation, but also other G proteins and further ion channels can be activated. This feature makes next level in muscarinic receptors signaling complexity (Tobin, Giglio, and Lundgren 2009, Myslivecek, Farar, and Valuskova 2017, Mighiu and Heximer 2012).

2. M_4 MUSCARINIC RECEPTORS

2.1. M_4 Muscarinic Receptor Signaling Pathways

M_4 muscarinic receptors are coupled, as mentioned above, to $G_{i/o}$ G proteins (Mistry, Dowling, and Challiss 2005) and thus they are able to inhibit the adenylyl cyclase/cyclic AMP signaling pathway. As another level of signal pathways, M_4 muscarinic receptors are also able to couple with G_s (Nathanson 2000). Another aspect of this coupling is general fact, that both G_i- and G_q-coupled muscarinic receptors (although when expressed at high levels only), can increase cAMP levels by coupling to G_s and thus activate adenylyl cyclase. On the other hand, M_4 muscarinic receptors are spontaneously active and can cause constitutive inhibition of adenylyl cyclase (Migeon and Nathanson 1994). As mentioned earlier, M_4 muscarinic receptors are also able to affect ion channels. Thus, it was also shown that M_4 muscarinic receptors can enhance Ca^{2+}currents via the effect on Ca_{V1} channels (Hernández-Flores et al. 2014).

When considering the regulation of M_4 muscarinic receptor signaling, there are some regions in the promoter area of M_4 muscarinic receptor gene. For example, sequencing study on rat M_4 muscarinic receptor has shown the presence of cell type-specific silencer element in the promoter region (Mieda, Haga, and Saffen 1996). The expression of the M_4 muscarinic receptor is regulated by the neuron-restrictive silencer element/repressor element 1 (Mieda, Haga, and Saffen 1997). Promoter region also contains the cAMP response element (Migeon and Nathanson 1994). Other study identified that promoter region does not contain a TATA or CAAT box and has several consensus sequences for enhancer elements including five Sp-1 binding sites, one AP-2 site, one AP-3 binding site, and two E-boxes within the proximal 600 bp (Wood et al. 1995).

2.2. Sequencing Similarity between Muscarinic Receptor Subtypes as Nature of Limited Ability to Determine Specific Subtype

Muscarinic receptors reveal high sequencing similarity as discovered in last century using sequencing studies (Bonner et al. 1987, Peralta et al. 1987) what further led to revealing similarities in structure as described in crystallization studies (Thal et al. 2016, Haga et al. 2012). This fact is responsible for the limited selectivity of commonly used muscarinic receptor ligands. More muscarinic receptor subtypes are expressed in specific tissue what makes difficult to define muscarinic receptor subtype-specific roles. As discussed elsewhere (Myslivecek et al. 2008, Caulfield and Birdsall 1998, Dhein, van Koppen, and Brodde 2001, Doods et al. 1993, Doods et al. 1994, Choppin et al. 1999, Lazareno et al. 1998, Buckley, Hulme, and Birdsall 1990, Bolognesi et al. 1998, Wang, Shi, and Wang 2004), the selectivity of muscarinic receptor antagonists is limited. Similarly, there is the serious problem with muscarinic receptor subtypes antibodies. All these pitfalls will be discussed below. Important hopes have been given to knockout studies, but there are also some problems using this technique that are discussed further.

2.2.1. Radioligand Binding

The main problem, also mentioned above, in the identification of muscarinic receptor subtypes present in specific central nervous system area is the lack of high subtype-selective muscarinic antagonists. The muscarinic receptor subtypes affinities to some antagonists are shown in Table 1.

It can be deduced from this table that except mamba toxins like MT7, the antagonist selectivity is very poor, and the majority of ligands bind to more than one subtype with similar affinity.

Table 1. Antagonist affinity constants (log affinity or pKi values) for muscarinic receptor subtypes

	Receptor subtype				
Antagonist	M_1	M_2	M_3	M_4	M_5
Pirenzepine	7.8-8.5	6.3-6.7	6.7-7.1	7.1-8.1	6.2-7.1
Methoctramine	7.1–7.8	7.8–8.3	6.3–6.9	7.4–8.1	6.2–7.2
4-DAMP	8.6–9.2	7.8–8.4	8.9–9.3	8.4–9.4	8.9–9.0
Himbacine	7.0–7.2	8.0–8.3	6.9–7.4	8.0–8.8	6.1–6.3
AF-DX 116	5.8-6.9	7.1-7.3	5.5-6.6	6.2-7.0	5.4-6.6
AF-DX 384	7.3–7.5	8.2–9.0	7.2–7.8	8.0-8.7	6.3
DAU 5884	9.40 ± 0.04	7.40 ± 0.05	8.80 ± 0.03	8.50 ± 0.02	N.n.
Tripinamide	7.2-7.4	7.9-9.3	5.15-5.33	6.68-6.92	N.n.
Darifenacin	7.5–7.8	7.0–7.4	8.4–8.9	7.7–8.0	8.0–8.1
PD 102807	5.3-5.5	5.7-5.9	6.2-6.7	7.3-7.4	5.2-5.5
p-F-HHSiD	6.68-7.3	6.01-6.6	7.5-7.84	7.2	6.6-7.0
AQ-RA 741	7.6-7.8	8.21-8.9	7.4-7.5	7.9-8.2	5.8-6.1
MT7 toxin	9.8	<6	<6	<6	<6
MT2 toxin	6.49	4.7	4.7	6	5.7
MT3 toxin	7.1	<6	<6	8.7	<6

N.n. not known. Data were obtained from Caulfield and Birdsal, 1998; Dhein et al. 2001; Doods et al. 1993, Doods et al. 1994, Eglen and Nahorski, 2000; Choppin et al. 1999; Lazareno et al. 1998; Buckley et al. 1990b; Bolonesi et al. 1998; and Wang et al. 2004.

However, mamba toxins, in general, do bind irreversibly and some studies have confirmed its allosteric binding properties (Olianas et al. 2004, Bradley, Rowan, and Harvey 2003, Carsi, Valentine, and Potter 1999, Karlsson et al. 2000, Myslivecek et al. 2008) what make their use also problematic. In radioligand binding studies, it is, therefore, necessary to use a combination of various antagonists. Moreover, usually full competition curves should be derived from increasing antagonist concentration binding studies. This makes radioligand binding studies, that try to identify muscarinic receptor subtypes, very time and resources consuming. On the other hand, radioligand binding studies performed with careful competition biding provide the most accurate picture of muscarinic receptor subtype distribution in respective tissue.

Another possibility how to determine muscarinic receptor subtypes in the specific tissue is the use of tritiated relatively specific ligands like ^{3}H-pirenzepine, ^{3}H-AFDX-384, ^{3}H-(+)telenzepine, ^{3}H-oxotremorine-M, ^{3}H-darifenacin (Alexander et al. 2017).

2.2.2. *Autoradiography*

Muscarinic receptors in the central nervous system can be detected either by radioligand binding studies or by autoradiography. Autoradiography is not a single method but refers to a concept used by a family of experimental techniques. The principle of autoradiography is the visualization and quantification of the radioactive substance within the specimen. In principle, the receptor binding of radioactive substance to specific target is not different from radioligand binding (Farar and Myslivecek 2016).

The total number of muscarinic receptors only can be detected by autoradiography. Muscarinic receptor subtypes cannot be detected by competition with antagonist because of evaluation limitations of such changed "binding" (Farar and Myslivecek 2016). Thus, some muscarinic receptor subtypes (M_1 and M_2 muscarinic receptor) have been detected in the central nervous system using tritiated ligands - ^{3}H-pirenzepine and ^{3}H-AFDX-384. Due to the limited selectivity of these ligands (see Table 1) the subtypes identification should be considered as the method for detection of M_1 (or M_2) and some portion of M_4 muscarinic receptors. Unfortunately, only a few papers report these binding sites as M_1/M_4 muscarinic receptors (Zavitsanou et al. 2003, Wang et al. 2014).

Historically, tritiated pirenzepine was used as a ligand that binds to muscarinic receptors with distinct binding in specific brain areas (Yamamura et al. 1983). Further, distinct distribution was found in several central nervous system regions what further led to the definition of binding sites as M_1 muscarinic receptors (Villiger and Faull 1985) and in the middle of eighties, pirenzepine binding sites are considered as M_1 muscarinic receptors (Cortes and Palacios 1986, Buckley and Burnstock 1986). On the other hand, AFDX-384 was from the beginning considered as M_2 muscarinic receptor-specific ligand (Aubert et al. 1992) and some authors became aware of limited selectivity (e.g., (Mulugeta et al. 2003)). In many cases, however, ^{3}H-AFDX-384 and ^{3}H-pirenzepine are considered as selective ligands (Tien et al. 2004, Wolff et al. 2008). We, therefore, performed series of autoradiographies with ^{3}H-AFDX-384, and ^{3}H-pirenzepine on M_1 muscarinic receptor knockout mice, M_2 muscarinic receptor knockout mice, and M_4 muscarinic receptor knockout mice to address the ligand selectivity (Valuskova, Farar, et al. 2018). We conclude that while ^{3}H-pirenzepine showed the high selectivity towards M_1 muscarinic receptors, ^{3}H-AFDX-384 binding sites represent different populations of muscarinic receptor subtypes in a brain and region-specific manner. This limited selectivity of radioligands should be considered when interpreting the data.

2.2.3. Western Blotting and Immunohistochemistry

Another possibility how to detect muscarinic receptor subtype is widely used methods that work with antibodies to specific muscarinic receptor subtypes - Western blotting and immunohistochemistry. Although these methods are used very often, the sequence similarity between muscarinic receptor subtypes leads to limited specific subtype detection (Pradidarcheep and Michel 2016). In detail, when verified by detection in knockout animals, only two antibodies (against M_2 muscarinic receptors) from 24 antibodies evaluated in 21 different protocol (i.e., 1,824 conditions evaluated) were specific (Jositsch et al. 2009). These findings were confirmed also in another study (Pradidarcheep et al. 2009). Thus, one should be aware of these limitations and specifically consider them when interpreting the results.

2.2.4. Gene Expression

Gene expression studies are also often the method used in muscarinic receptor subtypes determination (Jacoby et al. 1998, Hellgren et al. 2000, Benes, Varejkova, et al. 2012, Jackson and Nathanson 1995, Rousell et al. 1995, Jovanovic et al. 2018). Although the context of these findings is properly discussed in some papers, one should be aware that gene expression changes could not correlate with receptor subtype levels. This can be demonstrated on our study in which we determined mRNA levels and radioligand binding and shown different gene expression levels and muscarinic receptor levels in the heart (Myslivecek et al. 2006).

2.3. M_4 Muscarinic Receptor Function in the Central Nervous System

M_4 muscarinic receptors have a huge amount of functions like role in behavior (Bubser et al. 2014), social behavior (Koshimizu, Leiter, and Miyakawa 2012), learning and memory (Anagnostaras, Maren, and Fanselow 1995, Thomsen et al. 2012), attention (Chen, Baxter, and Rodefer 2004, Mirza and Stolerman 2000), locomotion (Sipos, Burchnell, and Galbicka 1999), thermoregulation, sleep and wakefulness (Sanford et al. 2006), food intake (Pratt and Blackstone 2009) and reward (Crespo et al. 2008).

The current understanding of the role of M_4 receptors comes from the use of genetically modified mice with either overall lack of M_4 MR or in individual neuronal subpopulations (Wess, Eglen, and Gautam 2007, Jeon et al. 2010). However, as mentioned above, gene targeting methods require special attention, especially when studying behavior.

M_4 muscarinic receptors can modify the neuronal circuits directly or indirectly. Direct (post-synaptic) effects are based on changes in neuronal excitability, transcription, and translation; indirect mechanisms are mediated through regulation of acetylcholine (ACh) tone by the regulation of neurotransmitter release (Zhang et al. 2002, Brown 2010). The direct regulation is mediated by postsynaptic muscarinic receptors, usually coupled to PLC such as M_1 muscarinic receptors. However, M_4 muscarinic receptors can provide postsynaptic inhibition by $G_{\beta\gamma}$ mediated direct activation of G-protein gated inwardly rectifying potassium channels (GIRK) (Brown 2010, Lüscher and Slesinger 2010). The increase neuronal excitability (by enhancing Ca^{2+}currents through Ca_{V1}-channels) is another possibility how to provide postsynaptic inhibition (Hernández-Flores et al. 2014).

Indirect effects (feedback control of ACh release) were proven by studies performed on M_4 muscarinic receptor knockout mice. In these animals, increased levels of ACh were shown in several brain regions, as determined by *in vivo* microdialysis (Tzavara et al. 2004, Tzavara et al. 2003). The regulation of ACh release might involve the inhibition of N and P/Q calcium channels via a membrane-delimited pathway, presumably by direct action of $G_{\beta\gamma}$ dimer on ion channels (Brown 2010, Yan and Surmeier 1996) as well as the activation of G protein-coupled inwardlyrectifying potassium channels (GIRK), resulting in hyperpolarization of cholinergic neurons (Calabresi et al. 1998, Bonsi et al. 2008). Another mechanism might lie downstream to the calcium entry and involve direct interaction with the exocytotic machinery (Blackmer et al. 2001, Kupchik et al. 2011).

The role of M_4 muscarinic receptors in neuronal excitability control is tightly connected with the function of M_1 muscarinic receptors. M_1 muscarinic receptors excite postsynaptic membrane by suppression of several potassium currents including the voltage and time-dependent K^+ current (I_M/KCNQ channel current), the Ca^{2+} activated K^+ current that generates after-hyperpolarization (I_{AHP}), a leak K^+ current (I_{leak}) (Brown 2010) and inwardly rectifying potassium channels K_{ir2} (Shen et al. 2007). Activation of M_1 muscarinic receptors lead also to the depolarization by increasing the mixed Na^+/K^+ hyperpolarization-activated current (I_h) and Ca^{2+} dependent nonspecific cation current (I_{cat}) (Fisahn et al. 2002).

The modulation of neurotransmitter release can affect long-term changes of synaptic strength (Bonsi et al. 2008, Wang et al. 2006) - the synaptic plasticity - which is considered to be the molecular mechanism underlying learning and memory (Malenka and Bear 2004).

Moreover, the presence of M_4 muscarinic receptors is not limited to cholinergic neurons only, but they can be also found as heteroreceptors at glutamatergic and GABAergic terminals (Hersch et al. 1994, Carr and Surmeier 2007, Koós and Tepper 2002) and dopamine neurons (Tzavara et al. 2003, Pancani et al. 2014). The interaction between M_4 muscarinic receptors and dopamine release is important for motor coordination. Especially, striatonigral pathway expresses both D_1 dopamine receptors and M_4 muscarinic receptors (Bernard, Levey, and Bloch 1999). Activation of D_1 dopamine receptors increases locomotor activity, whereas M_4 muscarinic receptors activation has opposite effects. This is also in agreement with the first knockout study on M_4 muscarinic receptors indicating enhanced D_1 dopamine receptor-mediated locomotor stimulation in M_4 KO mice (Gomeza et al. 1999). Other studies have also demonstrated the dependence of dopaminergic neurotransmission on M_4 muscarinic receptor function. M_4 KO mice had increased basal levels of dopamine and revealed enhanced dopamine responses to psychostimulants such as D-amphetamine and phencyclidine (PCP, angel dust) in the nucleus accumbens (Tzavara et al. 2004).

2.3.1. M_4 Muscarinic Receptors and Locomotion

The locomotor control, in general, includes the multilevel network of CNS areas including cortex, basal ganglia, thalamus, cerebellum, reticular formation, vestibular apparatus, and spinal medulla. In addition to that, locomotion can be influenced by CNS structures involved in biorhythm pacemakers with the aim to change the output according to the needs of the organism.

Wide range of receptors, including muscarinic/nicotinic acetylcholine receptors, dopamine receptors, GABA receptors, and excitatory amino acid receptors, are able to affect these circuits and balance between these receptors (and neurotransmitters targeting to receptors) is key factor for effective locomotor regulation (Tzavara et al. 2004, Gomeza et al. 1999, Carr and Surmeier 2007).

Although M_4 muscarinic receptors have been assigned to importantly affect the motor activity in the first knockout study (Gomeza et al. 1999), some contradictory findings have been reported (Turner, Hughes, and Toth 2010, Schmidt et al. 2011, Fink-Jensen et al. 2011, Woolley et al. 2009). These studies have reported the same locomotor activity in M_4 KO mice as in WT mice. The effects of M_4 muscarinic receptors have been assigned to the enhanced dopaminergic signaling at D_1 dopamine receptors (Gomeza et al. 1999). Also, a recent report has shown locomotor hyperactivity of M_4 KO mice in the open field and light/dark transition (Koshimizu, Leiter, and Miyakawa 2012). Although above-mentioned studies employed M_4 KO mice, there were substantial differences in experimental conditions that might account for different observations. Apart from differences in approach to measuring locomotor activity and gender of experimental mice, the genetic background of M_4 KO mice appears to be central to different outcomes. This could be important as we have shown recently that increased locomotion is apparent in the active (dark) phase of their diurnal cycle and in females only (Valuskova, Forczek, et al. 2018). This could be an important fact showing that the difference can be seen during the nocturnal phase only (i.e., in their active phase), probably thanks to the fact that mice are nocturnal animals.

Moreover, there are also differences in the genetic background in studied animals. The knockouts, that are hyperactive, were bred on mixed 129SvEv/CF-1 or pure 129SvEv genetic backgrounds (Koshimizu, Leiter, and Miyakawa 2012, Gomeza et al. 1999). On contrary, C57BL/6NTac or C57Bl/6J backgrounds lost the hyperactivity (Turner, Hughes, and Toth 2010, Schmidt et al. 2011, Fink-Jensen et al. 2011, Woolley et al. 2009). Our study, reporting the changes in motor activity in females only was performed on mice with the C57BL/6NTac genetic background.

2.4. The Pitfalls of Knockout Studies

Knockout studies were initially considered as the optimal method for detection of gene function (Bymaster et al. 2003, Stengel et al. 2000, Gomeza et al. 2001). However, some problems have appeared thanks to the intensive investigation in this field of research. Some of them have been listed above: the different genetic background could affect the behavior of mice with the deleted gene. For example, in the interesting paper studying another part of cholinergic system research (acetylcholinesterase knockout mice), it has been demonstrated, that the phenotype of gene-targeted animals depends on background genes and not solely on the mutations in the targeted gene (Duysen and Lockridge 2006).

Another important problem is the flanking allele effect (Crusio et al. 2009). Shortly, it is another, but the much more important problem, based on 'genetic background.' In detail, the genetic background consists of more alleles than those flanking a certain deleted gene but also those that could change the behavior of the mutated animal. These alleles are derived from the strain in which the mutation was induced. However, the flanking allele effect was not sometimes considered an important factor for behavior determination and animals were crossed as knockouts and mutants in two separate lines. It is therefore important to pay attention to the genetic background when interpreting the results, or when comparing the results from studies with inconsistency in genetic background. It is recommended to backcross the mice at least for 10 generations with aim to eliminate the flanking allele effect (Crusio et al. 2009).

Another factor, that also should be considered, is the effect on other systems that interact with the deleted gene. In general, it is clear, that deletion of one gene could and usually affect other physiological processes. For example, we have proven multiple times, that deletion of muscarinic receptor decreased the number of adrenoceptors, typically in the heart where the function of these receptors is opposite (Tomankova et al. 2015, Benes, Varejkova, et al. 2012, Benes, Novakova, et al. 2012). In this context, some knockouts that have been considered as life incompatible, like acetylcholinesterase deletion, have been shown to be viable, although with many limitations (Myslivecek, Duysen, and Lockridge 2007, Xie et al. 2000). Moreover, it has been also demonstrated that the sum of muscarinic receptor subtypes, determined using binding studies in knockout (M_1-M_5) animals in different brain regions tended to be greater than binding in WT mice (Oki et al. 2005). Thus, there could be some compensatory mechanisms in muscarinic receptor subtype number in muscarinic receptor knockout mice.

Thus, all these conditions should be taken into consideration when thinking about the knockout animals as models of one specific gene lacking.

2.5. M_4 Muscarinic Receptor in the Light of Biorhythm

Biorhythmic changes have been described for the first in ancient times. Further research identified suprachiasmatic nucleus as the main pacemaker that affects and can be affected by other peripheral pacemakers (Schroeder and Colwell 2013). From molecules, involved in cholinergic signaling, almost all reveal diurnal variation (Hut and Van der Zee 2011). While acetylcholinesterase, the main enzyme of acetylcholine catalysis, has the peak in the inactive phase, acetylcholine, as well as cholineacetyltransferase activity (the enzyme that synthesizes acetylcholine) have the peak of activity (level, respectively) in the active phase. Similarly to that, muscarinic receptor levels also reveal diurnal changes (Hut and Van der Zee 2011).

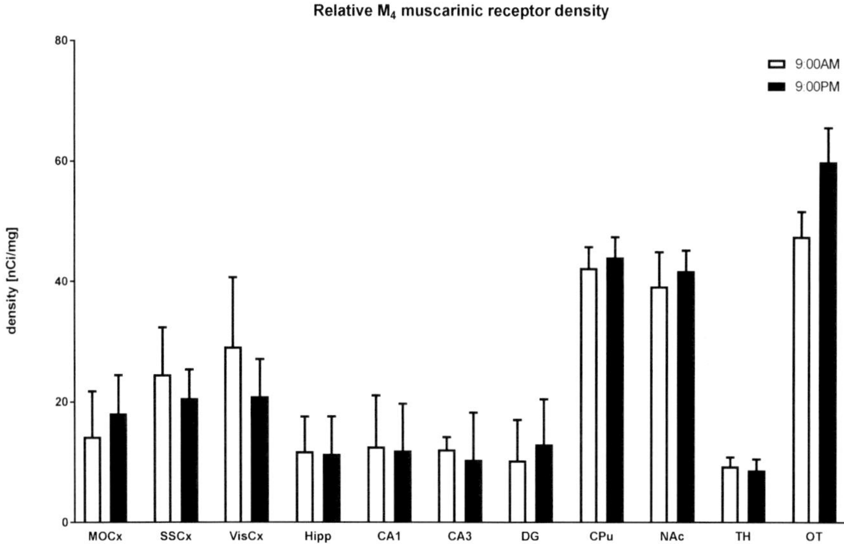

Figure 1. Relative M_4 muscarinic receptor densities in different brain areas. The results of autoradiography with ^{3}H-QNB (2 nM) in M_4 KO animals. Motor cortex (MOCx), somatosensory cortex (SSCx), visual cortex (VisCx), striatum (Caudatum-Putamen, CPu), nucleus accumbens (NAc), thalamus (TH), hippocampus (Hipp) and its specific areas CA1, CA3 and dentate gyrus (DG), olfactory tubercle (OT). No differences were found between 9:00 AM and 9:00 PM in any brain area. Data are shown as mean ± SEM.

It has also been shown that carbachol (unspecific muscarinic agonist) and McN-A-343 (relatively specific M_1/M_4 muscarinic agonist) but not bethanechol (relatively specific M_2/M_3 muscarinic agonist) caused phase advances when injected to the SCN during the mid-subjective day (Basu et al. 2016). On contrary, first results in muscarinic receptor directed changes in suprachiasmatic nucleus regulation of circadian rhythms have been ascribed to M_1 muscarinic receptors (Gillette et al. 2001). Concerning the mechanism, it seems probable that extra-suprachiasmatic cholinergic pathway is responsible for the phenomenon of carbachol-induced phase shift (Buchanan and Gillette 2005). We have compared densities of M_4 muscarinic receptors (using autoradiography in knockout animals) and did not found differences between 9:00 AM measurements and 9:00 PM measurements. However, there were substantial differences in densities in specific brain areas (see Figure 1).

However, it has been shown that muscarinic receptors (total number of muscarinic receptors, comprising all five subtypes M_1-M_5) variate diurnally in the visual cortex, but not in the suprachiasmatic nucleus (Bina, Rusak, and Wilkinson 1998).

CONCLUSION

All muscarinic receptor subtypes reveal relatively high amino acid sequence similarity and thus only a few muscarinic receptor ligands are selective. This makes the detection of MR in respective tissue problematic. All methods have pitfalls that should be considered in the interpretation of data. M_4 muscarinic receptors have been associated with various organism functions during the past years. Initially, their roles have been elucidated by means of pharmacological studies. These studies have indicated the important role of M_4 muscarinic receptors in several brain processes including learning and memory, attention, locomotion, thermoregulation, sleep and wakefulness, food intake and reward. It has been shown multiple times that M_4 muscarinic receptors regulate locomotion. However, there are some sex, genetic background, and biorhythm conditions consequences in this regulation.

Acknowledgment

Supported by Czech Science Foundation Grant N° 17-03847S.

References

Alexander, Stephen PH, Arthur Christopoulos, Anthony P Davenport, Eamonn Kelly, Neil V Marrion, John A Peters, Elena Faccenda, Simon D Harding, Adam J Pawson, Joanna L Sharman, Christopher Southan, and Jamie A Davies. 2017. "The Concise Guide to Pharmacology 2017/18: G protein-coupled receptors." *British Journal of Pharmacology* 174 (S1):S17-S129. doi: doi:10.1111/bph.13878.

Anagnostaras, S. G., S. Maren, and M. S. Fanselow. 1995. "Scopolamine selectively disrupts the acquisition of contextual fear conditioning in rats." *Neurobiology of Learning and Memory* 64 (3):191-194. doi: 10.1006/nlme.1995.0001.

Aubert, I., D. Cecyre, S. Gauthier, and R. Quirion. 1992. "Characterization and autoradiographic distribution of [3H]AF-DX 384 binding to putative muscarinic M2 receptors in the rat brain." *Eur J Pharmacol* 217 (2-3):173-84.

Basu, Priyoneel, Adrienne L. Wensel, Reid McKibbon, Nicole Lefebvre, and Michael C. Antle. 2016. "Activation of M1/4 receptors phase advances the hamster circadian clock during the day." *Neuroscience Letters* 621:22-27. doi: http://dx.doi.org/10.1016/j.neulet.2016.04.012.

Benes, Jan, Martina Novakova, Jana Rotkova, Vladimir Farar, Richard Kvetnansky, Vladimir Riljak, and Jaromir Myslivecek. 2012. "Beta3 Adrenoceptors Substitute the Role of M-2 Muscarinic Receptor in Coping with Cold Stress in the Heart: Evidence from M2KO Mice."

Cellular and Molecular Neurobiology 32 (5):859-869. doi: 10.1007/s10571-011-9781-3.

Benes, Jan, Eva Varejkova, Vladimir Farar, Martina Novakova, and Jaromir Myslivecek. 2012. "Decrease in heart adrenoceptor gene expression and receptor number as compensatory tool for preserved heart function and biological rhythm in M-2 KO animals." *Naunyn-Schmiedebergs Archives of Pharmacology* 385 (12):1161-1173. doi: 10.1007/s00210-012-0800-9.

Bernard, V., A. I. Levey, and B. Bloch. 1999. "Regulation of the subcellular distribution of m4 muscarinic acetylcholine receptors in striatal neurons in vivo by the cholinergic environment: evidence for regulation of cell surface receptors by endogenous and exogenous stimulation." *J Neurosci* 19 (23):10237-49.

Bina, K. G., B. Rusak, and M. Wilkinson. 1998. "Daily variation of muscarinic receptors in visual cortex but not suprachiasmatic nucleus of Syrian hamsters." *Brain Res* 797 (1):143-53.

Blackmer, T., E. C. Larsen, M. Takahashi, T. F. Martin, S. Alford, and H. E. Hamm. 2001. "G protein betagamma subunit-mediated presynaptic inhibition: regulation of exocytotic fusion downstream of Ca2+ entry." *Science* 292 (5515):293-297. doi: 10.1126/science.1058803.

Bolognesi, M. L., Minarini A., Budriesi R., Cacciaguerra S., Chiarini A., Spampinato S., Tumiatti V., and Melchiorre C. 1998. "Universal Template Aroach to Drug Design: Polyamines As Selective Muscarinic Receptor Antagonists." *J Med Chem* 41:4150-4160.

Bonner, T. I., N. J. Buckley, A. C. Young, and M. R. Brann. 1987. "Identification of a family of muscarinic acetylcholine receptor genes." *Science* 237 (4814):527-32.

Bonsi, Paola, Giuseppina Martella, Dario Cuomo, Paola Platania, Giuseppe Sciamanna, Giorgio Bernardi, Jürgen Wess, and Antonio Pisani. 2008. "Loss of muscarinic autoreceptor function impairs long-term depression but not long-term potentiation in the striatum." *J Neurosci* 28 (24):6258-6263. doi: 10.1523/JNEUROSCI.1678-08.2008.

Bradley, K. N., E. G. Rowan, and A. L. Harvey. 2003. "Effects of muscarinic toxins MT2 and MT7, from green mamba venom, on m1, m3 and m5 muscarinic receptors expressed in Chinese Hamster Ovary cells." *Toxicon* 41 (2):207-15.

Brown, David. 2010. "Muscarinic Acetylcholine Receptors (mAChRs) in the Nervous System: Some Functions and Mechanisms." *Journal of Molecular Neuroscience* 41 (3):340-346. doi: 10.1007/s12031-010-9377-2.

Bubser, Michael, Thomas M. Bridges, Ditte Dencker, Robert W. Gould, Michael Grannan, Meredith J. Noetzel, Atin Lamsal, Colleen M. Niswender, J. Scott Daniels, Michael S. Poslusney, Bruce J. Melancon, James C. Tarr, Frank W. Byers, Jürgen Wess, Mark E. Duggan, John Dunlop, Michael W. Wood, Nicholas J. Brandon, Michael R. Wood, Craig W. Lindsley, P. Jeffrey Conn, and Carrie K. Jones. 2014. "Selective Activation of M4 Muscarinic Acetylcholine Receptors Reverses MK-801-Induced Behavioral Impairments and Enhances Associative Learning in Rodents." *ACS Chemical Neuroscience* 5 (10):920-942. doi: 10.1021/cn500128b.

Buckley, N. J., E. C. Hulme, and N. J. Birdsall. 1990. "Use of clonal cell lines in the analysis of neurotransmitter receptor mechanisms and function." *Biochim Biophys Acta* 1055 (1):43-53.

Buckley, Noel J., and Geoffrey Burnstock. 1986. "Autoradiographic localization of peripheral M1 muscarinic receptors using [3H] pirenzepine." *Brain Research* 375 (1):83-91. doi: http://dx.doi.org/10.1016/0006-8993(86)90961-3.

Buchanan, Gordon F., and Martha U. Gillette. 2005. "New light on an old paradox: site-dependent effects of carbachol on circadian rhythms." *Experimental Neurology* 193 (2):489-496. doi: http://dx.doi.org/10.1016/j.expneurol.2005.01.008.

Bymaster, F. P., D. L. McKinzie, C. C. Felder, and J. Wess. 2003. "Use of M1-M5 muscarinic receptor knockout mice as novel tools to delineate the physiological roles of the muscarinic cholinergic system." *Neurochem Res* 28 (3-4):437-42.

Calabresi, P., D. Centonze, A. Pisani, G. Sancesario, R. A. North, and G. Bernardi. 1998. "Muscarinic IPSPs in rat striatal cholinergic interneurones." *Journal of physiology* 510 (Pt 2):421-427.

Carr, David B., and D. James Surmeier. 2007. "M1 muscarinic receptor modulation of Kir2 channels enhances temporal summation of excitatory synaptic potentials in prefrontal cortex pyramidal neurons." *Journal of neurophysiology* 97 (5):3432-3438. doi: 10.1152/jn. 00828.2006.

Carsi, J. M., H. H. Valentine, and L. T. Potter. 1999. "m2-toxin: A selective ligand for M2 muscarinic receptors." *Mol Pharmacol* 56 (5):933-7.

Caulfield, M. P., and N. J. Birdsall. 1998. "International Union of Pharmacology. XVII. Classification of muscarinic acetylcholine receptors." *Pharmacol Rev* 50 (2):279-90.

Chen, Katy C., Mark G. Baxter, and Joshua S. Rodefer. 2004. "Central blockade of muscarinic cholinergic receptors disrupts affective and attentional set-shifting." *European Journal of Neuroscience* 20 (4):1081-1088. doi: 10.1111/j.1460-9568.2004.03548.x.

Choppin, A., G. J. Stepan, D. N. Loury, N. Watson, and R. M. Eglen. 1999. "Characterization of the muscarinic receptor in isolated uterus of sham operated and ovariectomized rats." *Br J Pharmacol* 127 (7): 1551-8.

Cortes, R., and J. M. Palacios. 1986. "Muscarinic cholinergic receptor subtypes in the rat brain. I. Quantitative autoradiographic studies." *Brain Res* 362 (2):227-38.

Crespo, Jose A., Petra Stöckl, Katja Zorn, Alois Saria, and Gerald Zernig. 2008. "Nucleus accumbens core acetylcholine is preferentially activated during acquisition of drug- vs food-reinforced behavior." *Neuropsychopharmacology* 33 (13):3213-3220. doi: 10.1038/npp. 2008.48.

Crusio, W. E., D. Goldowitz, A. Holmes, and D. Wolfer. 2009. "Standards for the publication of mouse mutant studies." *Genes, Brain and Behavior* 8 (1):1-4.

Dhein, S., C. J. van Koppen, and O. E. Brodde. 2001. "Muscarinic receptors in the mammalian heart." *Pharmacol Res* 44 (3):161-82.

Doods, H. N., M. Entzeroth, H. Ziegler, N. Mayer, and P. Holzer. 1994. "Pharmacological profile of selective muscarinic receptor antagonists on guinea-pig ileal smooth muscle." *Eur J Pharmacol* 253 (3):275-81.

Doods, H. N., K. D. Willim, H. W. Boddeke, and M. Entzeroth. 1993. "Characterization of muscarinic receptors in guinea-pig uterus." *Eur J Pharmacol* 250 (2):223-30.

Farar, Vladimir, and Jaromir Myslivecek. 2016. "Autoradiography Assessment of Muscarinic Receptors in the Central Nervous System." In *Muscarinic Receptor: From Structure to Animal Models*, edited by Jaromir Myslivecek and Jan Jakubik, 159-180. Springer New York.

Fink-Jensen, Anders, Lene S. Schmidt, Ditte Dencker, Christina Schülein, Jürgen Wess, Gitta Wörtwein, and David P. D. Woldbye. 2011. "Antipsychotic-induced catalepsy is attenuated in mice lacking the M4 muscarinic acetylcholine receptor." *European Journal of Pharmacology* 656 (1–3):39-44. doi: http://dx.doi.org/10.1016/j.ejphar.2011.01.018.

Fisahn, André, Masahisa Yamada, Alokesh Duttaroy, Jai-Wei Gan, Chu-Xia Deng, Chris J. McBain, and Jürgen Wess. 2002. "Muscarinic induction of hippocampal gamma oscillations requires coupling of the M1 receptor to two mixed cation currents." *Neuron* 33 (4):615-624.

Gillette, M. U., G. F. Buchanan, L. Artinian, S. E. Hamilton, N. M. Nathanson, and C. Liu. 2001. "Role of the M1 receptor in regulating circadian rhythms." *Life Sci* 68 (22-23):2467-72.

Gomeza, J., L. Zhang, E. Kostenis, C. C. Felder, F. P. Bymaster, J. Brodkin, H. Shannon, B. Xia, A. Duttaroy, C. X. Deng, and J. Wess. 2001. "Generation and pharmacological analysis of M2 and M4 muscarinic receptor knockout mice." *Life Sciences* 68 (22-23):2457-66.

Gomeza, Jesus, Lu Zhang, Evi Kostenis, Christian Felder, Frank Bymaster, Jesse Brodkin, Harlan Shannon, Bing Xia, Chu-xia Deng, and Jürgen Wess. 1999. "Enhancement of D1 dopamine receptor-mediated locomotor stimulation in M4 muscarinic acetylcholine receptor knockout mice." *Proc Natl Acad Sci* 96 (18):10483-10488. doi: 10.1073/pnas.96.18.10483.

Haga, K., A. C. Kruse, H. Asada, T. Yurugi-Kobayashi, M. Shiroishi, C. Zhang, W. I. Weis, T. Okada, B. K. Kobilka, T. Haga, and T. Kobayashi. 2012. "Structure of the human M2 muscarinic acetylcholine receptor bound to an antagonist." *Nature* 482 (7386):547-51. doi: 10.1038/nature10753.

Hellgren, I., A. Mustafa, M. Riazi, I. Suliman, C. Sylven, and A. Adem. 2000. "Muscarinic M3 receptor subtype gene expression in the human heart." *Cell Mol Life Sci* 57 (1):175-80.

Hernández-Flores, Teresa, Omar Hernández-González, María B. Pérez-Ramírez, Esther Lara-González, Mario A. Arias-García, Mariana Duhne, Azucena Pérez-Burgos, Gilberto A. Prieto, Alejandra Figueroa, Elvira Galarraga, and José Bargas. 2014. "Modulation of direct pathway striatal projection neurons by muscarinic M4-type receptors." *Neuropharmacology*. doi: http://dx.doi.org/10.1016/j.neuropharm.2014.09.028.

Hersch, S. M., C. A. Gutekunst, H. D. Rees, C. J. Heilman, and A. I. Levey. 1994. "Distribution of m1-m4 muscarinic receptor proteins in the rat striatum: light and electron microscopic immunocytochemistry using subtype-specific antibodies." *Journal of Neuroscience* 14 (5 Pt 2):3351-3363.

Hut, R. A., and E. A. Van der Zee. 2011. "The cholinergic system, circadian rhythmicity, and time memory." *Behavioural Brain Research* 221 (2):466-480. doi: 10.1016/j.bbr.2010.11.039.

Jackson, D. A., and N. M. Nathanson. 1995. "Subtype-specific regulation of muscarinic receptor expression and function by heterologous receptor activation." *J Biol Chem* 270 (38):22374-7.

Jacoby, D. B., H. Q. Xiao, N. H. Lee, Y. Chan-Li, and A. D. Fryer. 1998. "Virus- and interferon-induced loss of inhibitory M2 muscarinic receptor function and gene expression in cultured airway parasympathetic neurons." *The Journal of Clinical Investigation* 102 (1):242-248. doi: 10.1172/JCI1114.

Jeon, Jongrye, Ditte Dencker, Gitta Wörtwein, David P. D. Woldbye, Yinghong Cui, Albert A. Davis, Allan I. Levey, Günther Schütz, Thomas N. Sager, Arne Mørk, Cuiling Li, Chu-Xia Deng, Anders

Fink-Jensen, and Jürgen Wess. 2010. "A Subpopulation of Neuronal M4 Muscarinic Acetylcholine Receptors Plays a Critical Role in Modulating Dopamine-Dependent Behaviors." *Journal of Neuroscience* 30 (6):2396-2405. doi: 10.1523/jneurosci.3843-09.2010.

Jositsch, Gitte, Tamara Papadakis, RainerV Haberberger, Miriam Wolff, Jürgen Wess, and Wolfgang Kummer. 2009. "Suitability of muscarinic acetylcholine receptor antibodies for immunohistochemistry evaluated on tissue sections of receptor gene-deficient mice." *Naunyn-Schmiedeberg's Archives of Pharmacology* 379 (4):389-395. doi: 10.1007/s00210-008-0365-9.

Jovanovic, Predrag, Natasa Spasojevic, Nela Puskas, Bojana Stefanovic, and Sladjana Dronjak. 2018. "Oxytocin modulates the expression of norepinephrine transporter, β3-adrenoceptors and muscarinic M2 receptors in the hearts of socially isolated rats." *Peptides*. doi: https://doi.org/10.1016/j.peptides.2018.06.008.

Karlsson, Evert, Mikael Jolkkonen, Ezra Mulugeta, Pierluigi Onali, and Abdu Adem. 2000. "Snake toxins with high selectivity for subtypes of muscarinic acetylcholine receptors." *Biochimie* 82 (9–10):793-806. doi: http://dx.doi.org/10.1016/S0300-9084(00)01176-7.

Koós, Tibor, and James M. Tepper. 2002. "Dual cholinergic control of fast-spiking interneurons in the neostriatum." *Journal of Neuroscience* 22 (2):529-535.

Koshimizu, Hisatsugu, Lorene Leiter, and Tsuyoshi Miyakawa. 2012. "M4 muscarinic receptor knockout mice display abnormal social behavior and decreased prepulse inhibition." *Molecular Brain* 5 (1):10.

Kupchik, Yonatan M., Ofra Barchad-Avitzur, Jürgen Wess, Yair Ben-Chaim, Itzchak Parnas, and Hanna Parnas. 2011. "A novel fast mechanism for GPCR-mediated signal transduction--control of neurotransmitter release." *Journal of Cell Biology* 192 (1):137-151. doi: 10.1083/jcb.201007053.

Lazareno, S., P. Gharagozloo, D. Kuonen, A. Popham, and N. J. Birdsall. 1998. "Subtype-selective positive cooperative interactions between brucine analogues and acetylcholine at muscarinic receptors: radioligand binding studies." *Mol Pharmacol* 53 (3):573-89.

Lüscher, Christian, and Paul A. Slesinger. 2010. "Emerging roles for G protein-gated inwardly rectifying potassium (GIRK) channels in health and disease." *Nature Reviews. Neuroscience* 11 (5):301-315. doi: 10.1038/nrn2834.

Malenka, Robert C., and Mark F. Bear. 2004. "LTP and LTD: an embarrassment of riches." *Neuron* 44 (1):5-21. doi: 10.1016/j.neuron.2004.09.012.

Mieda, M., T. Haga, and D. W. Saffen. 1996. "Promoter region of the rat m4 muscarinic acetylcholine receptor gene contains a cell type-specific silencer element." *Journal of Biological Chemistry* 271 (9):5177-5182.

Mieda, M., T. Haga, and D. W. Saffen. 1997. "Expression of the rat m4 muscarinic acetylcholine receptor gene is regulated by the neuron-restrictive silencer element/repressor element 1." *J Biol Chem* 272 (9):5854-60.

Migeon, J. C., and N. M. Nathanson. 1994. "Differential regulation of cAMP-mediated gene transcription by m1 and m4 muscarinic acetylcholine receptors. Preferential coupling of m4 receptors to Gi alpha-2." *J Biol Chem.* 269 (13):9767-73.

Mighiu, Alexandra, and Scott Patrick Heximer. 2012. "Controlling parasympathetic regulation of heart rate: a gatekeeper role for RGS proteins in the sinoatrial node." *Frontiers in Physiology* 3. doi: 10.3389/fphys.2012.00204.

Mirza, N. R., and I. P. Stolerman. 2000. "The role of nicotinic and muscarinic acetylcholine receptors in attention." *Psychopharmacology* 148 (3):243-250.

Mistry, Rajendra, Mark R. Dowling, and R. A. John Challiss. 2005. "An investigation of whether agonist-selective receptor conformations occur with respect to M2 and M4 muscarinic acetylcholine receptor signalling via Gi/o and Gs proteins." *British Journal of Pharmacology* 144 (4):566-575. doi: 10.1038/sj.bjp.0706090.

Mulugeta, Ezra, Evert Karlsson, Atiqul Islam, Raj Kalaria, Halinder Mangat, Bengt Winblad, and Abdu Adem. 2003. "Loss of muscarinic M4 receptors in hippocampus of Alzheimer patients." *Brain Research*

960 (1–2):259-262. doi: http://dx.doi.org/10.1016/S0006-8993(02)03542-4.

Myslivecek, J., E. G. Duysen, and O. Lockridge. 2007. "Adaptation to excess acetylcholine by downregulation of adrenoceptors and muscarinic receptors in lungs of acetylcholinesterase knockout mice." *Naunyn Schmiedebergs Arch Pharmacol* 376:83-92.

Myslivecek, J., V. Farar, and P. Valuskova. 2017. "M(4) muscarinic receptors and locomotor activity regulation." *Physiol Res* 66 (Supplementum 4):S443-s455.

Myslivecek, J., M. Novakova, M. Palkovits, O. Krizanova, and R. Kvetnansky. 2006. "Distribution of mRNA and binding sites of adrenoceptors and muscarinic receptors in the rat heart." *Life Sci* 79 (2):112-20.

Myslivecek, Jaromir, Martin Klein, Martina Novakova, and Jan Ricny. 2008. "The detection of the non-M-2 muscarinic receptor subtype in the rat heart atria and ventricles." *Naunyn-Schmiedebergs Archives of Pharmacology* 378 (1):103-116. doi: 10.1007/s00210-008-0285-8.

Nathanson, N. M. 2000. "A multiplicity of muscarinic mechanisms: enough signaling pathways to take your breath away." *Proc Natl Acad Sci* 97 (12):6245-7. doi: 10.1073/pnas.97.12.6245.

Oki, Tomomi, Yukiko Takagi, Sayuri Inagaki, Makoto M. Taketo, Toshiya Manabe, Minoru Matsui, and Shizuo Yamada. 2005. "Quantitative analysis of binding parameters of [3H]N-methylscopolamine in central nervous system of muscarinic acetylcholine receptor knockout mice." *Molecular Brain Research* 133 (1):6-11.

Olianas, M. C., A. Adem, E. Karlsson, and P. Onali. 2004. "Action of the muscarinic toxin MT7 on agonist-bound muscarinic M1 receptors." *Eur J Pharmacol* 487 (1-3):65-72.

Pancani, Tristano, Caroline Bolarinwa, Yoland Smith, Craig W. Lindsley, P. Jeffrey Conn, and Zixiu Xiang. 2014. "M4 mAChR-mediated modulation of glutamatergic transmission at corticostriatal synapses." *ACS Chemical Neuroscience*. doi: 10.1021/cn500003z.

Peralta, E. G., A. Ashkenazi, J. W. Winslow, D. H. Smith, J. Ramachandran, and D. J. Capon. 1987. "Distinct primary structures, ligand-binding properties and tissue-specific expression of four human muscarinic acetylcholine receptors." *Embo j* 6 (13):3923-9.

Pradidarcheep, Wisuit, and Martin C. Michel. 2016. "Use of Antibodies in the Research on Muscarinic Receptor Subtypes." In *Muscarinic Receptor: From Structure to Animal Models*, edited by Jaromir Myslivecek and Jan Jakubik, 83-94. New York, NY: Springer New York.

Pradidarcheep, Wisuit, Jan Stallen, Wil Labruyère, Noshir Dabhoiwala, Martin Michel, and Wouter Lamers. 2009. "Lack of specificity of commercially available antisera against muscarinergic and adrenergic receptors." *Naunyn-Schmiedeberg's Archives of Pharmacology* 379 (4):397-402.

Pratt, Wayne E., and Kaitlin Blackstone. 2009. "Nucleus accumbens acetylcholine and food intake: decreased muscarinic tone reduces feeding but not food-seeking." *Behavioural Brain Research* 198 (1):252-257. doi: 10.1016/j.bbr.2008.11.008.

Rousell, J., E. B. Haddad, J. C. W. Mak, and P. J. Barnes. 1995. "Transcriptional down-regulation of m2 muscarinic receptor gene expression in human embryonic lung (HEL 299) cells by protein kinase C." *Journal of Biological Chemistry* 270 (13):7213-7218.

Sanford, L. D., L. Yang, X. Tang, E. Dong, R. J. Ross, and A. R. Morrison. 2006. "Cholinergic regulation of the central nucleus of the amygdala in rats: effects of local microinjections of cholinomimetics and cholinergic antagonists on arousal and sleep." *Neuroscience* 141 (4):2167-2176. doi: 10.1016/j.neuroscience.2006.05.064.

Shen, Weixing, Xinyong Tian, Michelle Day, Sasha Ulrich, Tatiana Tkatch, Neil M. Nathanson, and D. James Surmeier. 2007. "Cholinergic modulation of Kir2 channels selectively elevates dendritic excitability in striatopallidal neurons." *Nature Neuroscience* 10 (11):1458-1466. doi: 10.1038/nn1972.

Schmidt, Lene S, Morgane Thomsen, Pia Weikop, Ditte Dencker, Jürgen Wess, David P D. Woldbye, Gitta Wortwein, and Anders Fink-Jensen. 2011. "Increased cocaine self-administration in M4 muscarinic acetylcholine receptor knockout mice." *Psychopharmacology* 216 (3):367-378. doi: 10.1007/s00213-011-2225-4.

Schroeder, A. M., and C. S. Colwell. 2013. "How to fix a broken clock." *Trends Pharmacol Sci* 34 (11):605-619. doi: 10.1016/j.tips.2013. 09.002.

Sipos, M. L., V. Burchnell, and G. Galbicka. 1999. "Dose-response curves and time-course effects of selected anticholinergics on locomotor activity in rats." *Psychopharmacology* 147 (3):250-256.

Stengel, P. W., J. Gomeza, J. Wess, and M. L. Cohen. 2000. "M(2) and M(4) receptor knockout mice: muscarinic receptor function in cardiac and smooth muscle in vitro." *J Pharmacol Exp Ther* 292 (3):877-85.

Thal, David M., Bingfa Sun, Dan Feng, Vindhya Nawaratne, Katie Leach, Christian C. Felder, Mark G. Bures, David A. Evans, William I. Weis, Priti Bachhawat, Tong Sun Kobilka, Patrick M. Sexton, Brian K. Kobilka, and Arthur Christopoulos. 2016. "Crystal structures of the M1 and M4 muscarinic acetylcholine receptors." *Nature* 531:335-340.

Thomsen, Morgane, Craig W Lindsley, P. Jeffrey Conn, Jeffrey E Wessell, BrianS Fulton, Jürgen Wess, and S. Barak Caine. 2012. "Contribution of both M1 and M4 receptors to muscarinic agonist-mediated attenuation of the cocaine discriminative stimulus in mice." *Psychopharmacology* 220 (4):673-685. doi: 10.1007/s00213-011-2516-9.

Tien, Lu-Tai, Lir-Wan Fan, Chiharu Sogawa, Tangeng Ma, Horance H. Loh, and Ing-Kang Ho. 2004. "Changes in acetylcholinesterase activity and muscarinic receptor bindings in μ-opioid receptor knockout mice." *Molecular Brain Research* 126 (1):38-44. doi: http://dx.doi.org/10. 1016/j.molbrainres.2004.03.011.

Tobin, G., D. Giglio, and O. Lundgren. 2009. "Muscarinic receptor subtypes in the alimentary tract." *J Physiol Pharmacol* 60 (1):3-21.

Tomankova, Hana, Paulina Valuskova, Eva Varejkova, Jana Rotkova, Jan Benes, and Jaromir Myslivecek. 2015. "The M 2 muscarinic receptors are essential for signaling in the heart left ventricle during restraint stress in mice." *Stress* 18 (2):208-20. doi: 10.3109/10253890. 2015.1007345.

Turner, Jeremy, Larry F. Hughes, and Linda A. Toth. 2010. "Sleep, activity, temperature and arousal responses of mice deficient for muscarinic receptor M2 or M4." *Life Sciences* 86 (5–6):158-169. doi: http://dx.doi.org/10.1016/j.lfs.2009.11.019.

Tzavara, E. T., F. P. Bymaster, C. C. Felder, M. Wade, J. Gomeza, J. Wess, D. L. McKinzie, and G. G. Nomikos. 2003. "Dysregulated hippocampal acetylcholine neurotransmission and impaired cognition in M2, M4 and M2/M4 muscarinic receptor knockout mice." *Mol Psychiatry.* 8 (7):673-9.

Tzavara, Eleni T., Frank P. Bymaster, Richard J. Davis, Mark R. Wade, Kenneth W. Perry, Jurgen Wess, David L. McKinzie, Chris Felder, and George G. Nomikos. 2004. "M4 muscarinic receptors regulate the dynamics of cholinergic and dopaminergic neurotransmission: relevance to the pathophysiology and treatment of related CNS pathologies." *FASEB journal* 18 (12):1410-1412. doi: 10.1096/fj.04-1575fje.

Valuskova, P., S. T. Forczek, V. Farar, and J. Myslivecek. 2018. "The deletion of M4 muscarinic receptors increases motor activity in females in the dark phase." *Brain Behav* 8 (8):e01057. doi: 10.1002/brb3.1057.

Valuskova, Paulina, Vladimir Farar, Sandor Forczek, Iva Krizova, and Jaromir Myslivecek. 2018. "Autoradiography of 3H-pirenzepine and 3H-AFDX-384 in Mouse Brain Regions: Possible Insights into M1, M2, and M4 Muscarinic Receptors Distribution." *Frontiers in Pharmacology* 9 (124). doi: 10.3389/fphar.2018.00124.

Villiger, John W., and Richard L. M. Faull. 1985. "Muscarinic cholinergic receptors in the human spinal cord: differential localization of [3H]pirenzepine and [3H]quinuclidinylbenzilate binding sites." *Brain Research* 345 (1):196-199. doi: http://dx.doi.org/10.1016/0006-8993(85)90854-6.

Wang, Q., X. Wei, H. Gao, J. Li, J. Liao, X. Liu, B. Qin, Y. Yu, C. Deng, B. Tang, and X. F. Huang. 2014. "Simvastatin reverses the downregulation of M1/4 receptor binding in 6-hydroxydopamine-induced parkinsonian rats: The association with improvements in long-term memory." *Neuroscience* 267 (0):57-66. doi: http://dx.doi.org/10.1016/j.neuroscience.2014.02.031.

Wang, Z., H. Shi, and H. Wang. 2004. "Functional M3 muscarinic acetylcholine receptors in mammalian hearts." *Br J Pharmacol* 142 (3):395-408.

Wang, Zhongfeng, Li Kai, Michelle Day, Jennifer Ronesi, Henry H. Yin, Jun Ding, Tatiana Tkatch, David M. Lovinger, and D. James Surmeier. 2006. "Dopaminergic control of corticostriatal long-term synaptic depression in medium spiny neurons is mediated by cholinergic interneurons." *Neuron* 50 (3):443-452. doi: 10.1016/j.neuron.2006.04.010.

Wess, J., R. M. Eglen, and D. Gautam. 2007. "Muscarinic acetylcholine receptors: mutant mice provide new insights for drug development." *Nat Rev Drug Discov* 6 (9):721-33.

Wolff, Samuel C., Zuzana Hruska, Lieuko Nguyen, and Gary P. Dohanich. 2008. "Asymmetrical distributions of muscarinic receptor binding in the hippocampus of female rats." *European Journal of Pharmacology* 588 (2–3):248-250. doi: http://dx.doi.org/10.1016/j.ejphar.2008.04.002.

Wood, I. C., A. Roopra, C. Harrington, and N. J. Buckley. 1995. "Structure of the m4 cholinergic muscarinic receptor gene and its promoter." *Journal of Biological Chemistry* 270 (52):30933-30940.

Woolley, Marie L., Helen J. Carter, Jane E. Gartlon, Jeanette M. Watson, and Lee A. Dawson. 2009. "Attenuation of amphetamine-induced activity by the non-selective muscarinic receptor agonist, xanomeline, is absent in muscarinic M4 receptor knockout mice and attenuated in muscarinic M1 receptor knockout mice." *European Journal of Pharmacology* 603 (1–3):147-149. doi: http://dx.doi.org/10.1016/j.ejphar.2008.12.020.

Xie, W., J. A. Stribley, A. Chatonnet, P. J. Wilder, A. Rizzino, R. D. McComb, P. Taylor, S. H. Hinrichs, and O. Lockridge. 2000. "Postnatal developmental delay and supersensitivity to organophosphate in gene-targeted mice lacking acetylcholinesterase." *J Pharmacol Exp Ther* 293 (3):896-902.

Yamamura, Henry I., James K. Wamsley, Pushpa Deshmukh, and William R. Roeske. 1983. "Differential light microscopic autoradiographic localization of muscarinic cholinergic receptors in the brainstem and spinal cord of the rat using [3H]pirenzepine." *European Journal of Pharmacology* 91 (1):147-149. doi: http://dx.doi.org/10.1016/0014-2999(83)90379-5.

Yan, Z., and D. J. Surmeier. 1996. "Muscarinic (m2/m4) receptors reduce N- and P-type Ca2+ currents in rat neostriatal cholinergic interneurons through a fast, membrane-delimited, G-protein pathway." *Journal of Neuroscience* 16 (8):2592-2604.

Zavitsanou, Katerina, Andrew Katsifis, Filomena Mattner, and Xu-Feng Huang. 2003. "Investigation of M1//M4 Muscarinic Receptors in the Anterior Cingulate Cortex in Schizophrenia, Bipolar Disorder, and Major Depression Disorder." *Neuropsychopharmacology* 29 (3):619-625.

Zhang, Weilie, Anthony S. Basile, Jesus Gomeza, Laura A. Volpicelli, Allan I. Levey, and Jürgen Wess. 2002. "Characterization of Central Inhibitory Muscarinic Autoreceptors by the Use of Muscarinic Acetylcholine Receptor Knock-Out Mice." *Journal of Neuroscience* 22 (5):1709-1717.

In: Acetylcholine Receptors …
Editor: Adelais Eros Gupta
ISBN: 978-1-53615-447-4

Chapter 3

Nicotine and Agonists of Alpha 7 Nicotinic Acetylcholine Receptors Combined with Environmental Enrichment

Patricia Mesa-Gresa, PhD and Rosa Redolat*, PhD
Department of Psychobiology, Universitat de València,
Valencia, Spain

Abstract

Nicotine and agonists of alpha7 nicotinic acetylcholine receptors (nAChRs) are being evaluated as a possible treatment for cognitive symptoms associated to schizophrenia, Alzheimer's disease and other neuropscyhiatric and neurodegenerative disorders. It has been proposed that these drugs can act as cognitive enhancers, improving memory and learning in different tasks. Therefore, it could be of interest to evaluate their effects in preclinical studies in which different drugs are administered in conjunction with other therapeutic strategies such as environmental interventions based on complex environments including

* Corresponding Author's E-mail: Rosa.Redolat@uv.es.

higher social, cognitive and physical stimulation than standard environments. Prior studies in rodents suggest that these enriched environments induce behavioral and neurobiological changes which can be related to diminished anxiety-like behavior and an improvement in the performance of learning and memory tasks. In the present chapter, we will review the environmental enrichment (EE) paradigm in rodents and their effects on neurobiological, physiological and behavioral variables in preclinical studies. We will also present results obtained in our laboratory regarding effects of nicotine and PNU 282987 in interaction with EE in different behavioral tasks. We will conclude indicating the implications that experimental results may have for the search and identification of new therapeutic approaches aimed to counteract or delay cognitive deficits observed in neurodegenerative or neuropsychiatric disorders.

Keywords: nicotinic acetylcholine receptors, environmental enrichment, cognition, Alzheimer's disease, mental illness, rodents

INTRODUCTION

Environmental enrichment (EE) has been proposed as a therapeutic intervention which may induce affective and cognitive changes (Redolat and Mesa-Gresa, 2012; Rogers et al., 2017; Sale, 2018). These beneficial effects of complex environments including social, cognitive and somatosensorial stimulation have been demonstrated in experimental studies conducted both in rodents (Pang and Hannan, 2013) and in human subjects (Sale, 2018). The neurobiological and behavioral changes induced by environmental manipulations have been recently proposed as an interesting approach in the advancement of knowledge about experience-induced brain plasticity as well as a useful intervention aimed to the development and testing of new treatments against cognitive decline associated with aging, neurodegenerative pathologies and mental disorders (Leon and Woo, 2018; Prado-Lima et al., 2017; Sampedro-Piquero and Begega, 2017).

In the present chapter we will discuss how a pharmacological treatment (administration of nicotine or agonists of nicotinic alpha 7 receptors) could be modulated by the exposure to complex and stimulant environments. We will focus on the main effects induced by EE both at neurobiological and behavioral levels. Furthermore, we will describe the possible use of nicotine and nicotinic agonists for the treatment of cognitive symptoms associated to schizophrenia or Alzheimer's disease. The research about the use of pharmacological treatments based on nicotine or nicotinic alpha 7 agonists is a topic of interest due to the lack of effective treatments for cognitive symptoms.

Finally, we will conclude with a translational perspective from the studies reviewed in order to obtain a broader picture about changes in cognition and neural plasticity related to aging and neurodegenerative disorders in human subjects.

NICOTINE AND ALPHA 7 NICOTINIC ACETYLCHOLINE RECEPTORS

Nicotine and agonists of alpha7 nicotinic acetylcholine receptors (nAChRs) are being evaluated as a possible treatment for different mental and neurodegenerative disorders (Bertrand and Terry, 2018; Jones, 2018; Kucinski et al., 2011; Lombardo and Maskos, 2015; Yang et al., 2017). It has been proposed that these drugs can act as cognitive enhancers, improving learning and memory in different tasks (Carrasco et al., 2006; Hsu et al., 2018; Valentine and Sofuoglu, 2018). The cognitive enhancement effects of nicotinic agonists have also been supported in different experimental studies with rodents (Freedman, 2014; Jones et al., 2014; McLean et al., 2011, 2016) although more research is needed on human subjects in order to establish the main effects of these drugs in neurodegenerative and mental diseases (Jones, 2018).

The search for a treatment for Alzheimer's disease continues and the need to identify a treatment addressing both cognitive and behavioral symptoms has been underscored (Lanctôt et al., 2017; Pfeffer et al., 2018). Recently, it has been proposed that a more global perspective is needed in order to find more effective treatments for this neurodegenerative disorder (Gravitz, 2018).

A review about the possible use of nicotine and nicotine agonists for the treatment of Alzheimer's disease concluded that α7 is the primary nAChR subtype that is responsible for cognition and memory and these receptors have been the major recent experimental targets for nAChRs agonist strategy. Different nicotinic agonists (AZD0328, SSR180711, RG3487, TC-5619…) have been evaluated in the search of new treatment strategies for Alzheimer's disease or other neurodegenerative disorders (Verma et al., 2018).

Animal studies are needed to model all the main clinical symptoms of Alzheimer's disease. These experimental approaches should display the key neuropathological changes of this neurodegenerative disorder, including not only cognitive deficits but also psychological and behavioral symptoms. This exhaustive characterization will allow the progress of the knowledge in the search of more effective and personalized treatments for this disease (Kossik, 2015). An interesting in depth review including all rodent models currently available for Alzheimer's disease can be found in Götz et al., (2018). It has been suggested that is important to develop a standardized model and common protocols which could be applied in different laboratories (Götz et al., 2018). In relation to Parkinson's disease, the need to take into account both motor and non-motor symptoms (cognitive, neuropsychiatric...) has also recently been emphasized in order to identify new treatments for this disorder (Titova and Chaudhuri, 2018).

There is no effective treatment for cognitive symptoms in schizophrenia and for this reason the need to explore this topic has recently been highlighted (Hsu et al., 2018). EVP-6124 (Encenicline) induced cognitive improvement in schizophrenia patients and arrived to phase 3 trial but was withdrawn from research due to gastrointestinal secondary effects (Verma et al., 2018). In animal models for schizophrenia, the need to advance in modelling cognitive alterations that are currently considered key in the search for effective treatments for this mental disorder has also been pointed out (Nikiforuk, 2018). Some recommendations to enhance the translational value of animal models have been formulated by *Cognitive*

Neuroscience Treatment Research to Improve Cognition in Schizophrenia (CINTRICS) (Nikiforuk, 2018). New models based on technological approaches (such as preclinical models using "touch screen") could help in this identification of more useful models to evaluate therapeutic interventions.

In the present review we propose that it could be of interest to evaluate the effects of nicotinic agonists when they are administered in conjunction with other therapeutic strategies such as enrichment of the environment. These preclinical studies could contribute to identify new perspectives in the treatment of these mental and neurodegenerative disorders including not only pharmacological approaches but also behavioral ones, obtaining multimodal interventions (Rogers et al., 2017). In fact, interventions based in lifestyle changes are being proposed as an interesting approach in the prevention of neurodegenerative diseases.

The Environmental Enrichment Paradigm

The EE paradigm has been widely studied as a model to investigate the effects of the exposure to complex conditions and the impact of environmental experience on the brain and its behavioral correlates (Redolat and Mesa-Gresa, 2012; Sale, 2018). This type of allocation is based on exposure of laboratory animals to an environment that includes cognitive, social, physical and sensorial stimulation and a greater opportunity to explore, to improve cognitive function, to engage in social interaction and to display physical activity (Redolat and Mesa-Gresa, 2012; Hannan, 2014; Sale et al., 2014).

Different procedures have been carried out from diverse laboratories in order to expose the animals to EE conditions. Main components include larger cages that allow the allocation of more animals than in standard conditions (4 to 20 animals), inanimate objects with different textures and materials that could be removed and changed constantly (toys, balls, houses, construction pieces, mirrors, bells, etc.), and other objects that increase physical activity of the rodens (ladders, running wheels, ropes, platforms, etc…). All these objects can be combined into different protocols that increase the exploration and curiosity of animals through a novel, complex and challenging environment (Mo et al., 2016; Mesa-Gresa et al., 2016). Novelty and complexity seem to be the key features related to plastic changes in the brain induced by EE (Redolat and Mesa-Gresa, 2012). Animals could be exposed to EE before birth (prenatal enrichment models), during early ages or into adulthood, including different periods and methods of exposure. Significative effects have been reported in animals exposed to EE during different periods of time (from 24h to several months), age of exposure and frequency (some hours per day or all day). The presence or absence of some of the objects described may also induce differences between results obtained in different laboratories (Hannan, 2014). In our laboratory we have carried out different studies in which the influence of each one of the components of the EE in the results obtained has been analyzed in detail (Mesa-Gresa et al., 2013a).

There is no consensus about a specific protocol for EE exposure in rodents. Furthermore, we can found in the literature a lot of differences between laboratories. The lack of a standardized protocol of EE could explain some discrepancies in the results obtained in the published studies. Recently, new perspectives have been established in order to define more specifically the methodological points that should be included in the development of an EE model (Sale, 2018). The lack of a standardized protocol regarding the application of the EE paradigm in rodents makes it difficult to establish the main benefits and/or effects of exposure to enriched housing conditions. Main differences between laboratories are related to the dimensions of the cages, number of animals per cage, objects included, the total period of exposure, the complexity of the environment or the frequency of the exposition and/or the objects' changes (Redolat and Mesa-Gresa, 2012; Mesa-Gresa et al., 2014).

Main Consequences of Exposure to Environmental Enrichment in Rodents

Main neurobiological and behavioral effects of the exposure to enriched environmental conditions have been analyzed in the scientific literature. In spite of the differences between different EE procedures previously explained, there is evidence about the brain and behavioral impact of environmental manipulations, obtaining significant effects in both enriched and impoverished environments. Taking into account the main aim of the present chapter, we will review the physical, neurobiological and behavioral effects induced by EE in laboratory animals.

Regarding physical effects of EE, there are evidence about changes in body weight gain and in food or fluid intake. Animals exposed to EE display in general lower weight gain and higher food and fluid intake than animals maintained in standard conditions (Mesa-Gresa et al., 2013b, 2016). These results could be related to the physical changes induced by the complex environments as well as to greater opportunities of movement, jumping and running provided by the different objects contained in the larger enriched cages.

Exposure to EE also has an important neurobiological impact. Diverse studies have shown significant structural and functional changes related to EE exposure in rodents. Main morphological consequences included an increase in cortical thickness and volume, dendritic branching and spines, cell survival, synaptic connection and size of neuronal somas. Other interesting neurobiological effects of EE include an enhancement of the synaptic plasticity, neurotransmitter levels, number of receptors, neurogenesis in hippocampal areas and increase in brain-derived neurotrophic factors (Alwis and Rajan, 2014; Nithianantharajah and Hannan, 2011; Sale et al., 2014; Sale, 2018; Sampedro-Piquero and Begega, 2017). Furthermore, changes in cognitive reserve have been proposed as a consequence of this exposure to enriched environments

(Redolat and Mesa-Gresa, 2012). The effects of the EE in the brain could also induce alterations in neuroendocrine response, obtaining differences in the activation, for instance, of the hipothalamic-piyuitary-adrenal (HPA) axis. Animals living in enriched environments showed lower activation of HPA axis but there are not conclusive results about corticosterone levels (Mesa-Gresa et al., 2016; Sampedro-Piquero and Begega, 2017).

All these neurobiological changes could have behavioral correlates. Diminished motor and exploratory behavior in animals exposed to EE has been reported. Different studies have also confirmed a significant improvement in the performance of learning and memory tasks, especially in hippocampal dependent tasks (Harati et al., 2011; Mesa-Gresa et al., 2013b, 2013c; Sale et al., 2014). Regarding the emotional response, a decrease in anxiety-like levels measured in the elevated plus-maze and in other animal models has been reported (Mesa-Gresa et al., 2013b). Research performed in different laboratories has also indicated that EE could counteract the effects of exposure to stress and increase the resistance to different type of stressors (Dandi et al., 2018; Mesa-Gresa et al., 2016; Novaes et al., 2017). Other results indicate that EE conditions could have antidepressant effects in animal models of depression (Seong et al., 2018). Other behavioral parameters related with social interaction, as aggressive behavior, have received less attention and data are not conclusive (Mesa-Gresa et al., 2013c).

The EE paradigm has been applied in animals at different ages, obtaining positive effects in cognitive function and in the prevention of age-related cognitive impairment. These results may be of interest for the prevention and/or treatment of neurodegenerative disorders such as Alzheimer's disease (Prado Lima et al., 2018), Parkinson's disease (Jungling et al., 2017), Huntington's disease (Mo et al., 2015), and other pathologies related with cognitive impairment as brain injury (de la Tremblaye et al., 2018; Gonçalves et al., 2018) or schizophrenia (Bator et al., 2018; Burrows and Hannan, 2016; Tregellas and Wylie, 2018, Young and Geyer, 2013).

Combination of Environmental Enrichment and Administration of Agonists of Alpha 7 Nicotinic Acetylcholine Receptors

The environment also plays an important role in the development of drug addiction. Social influences, lifestyle and the level of neurocognitive stimulation, among other factors, have proved to be especially important in predicting drug use. As discussed in the previous sections, complexity and enrichment of the environment give rise to neurochemical, cellular and molecular changes that could play a fundamental role in addiction processes (Bockman et al., 2018; Piña et al., 2018). For that reason, different studies have recently highlighted the need of evaluating the effects of the environment on the development of addiction to different drugs such as nicotine, heroin or metamphetamines, including the assessment of social influences (Piña et al., 2018; Sikora et al., 2018; Stairs et al., 2016). Preclinical studies can help us to obtain information about the consumption of some drugs, as well as about the therapeutic possibilities offered by substances such as nicotine or nicotinic agonists in the treatment of neurodegenerative or mental diseases.

Few studies have evaluated the effects of EE in combination with pharmacological manipulations such as nicotine or other agonists of nAChRs. A review of these studies can be found in Mesa-Gresa et al. (2013a) and Stairs et al. (2016). Both manipulations (pharmacological and non-pharmacological) have demonstrated significant effects on the functions of attention, memory and learning (Mesa-Gresa et al., 2014; Levin, 2012; Stairs et al., 2016). It has been suggested that a summative consequence of the cognitive effects of EE and the administration of cholinergic agonists may be produced.

Current evidence indicates that the allocation of rodents in enriched conditions can induce neurobiological effects related to dopamine release on prefrontal cortex and nucleus accumbens (Zhu et al., 2007; 2013) and hormonal release induced by nicotine administration (Skwara et al., 2012). Behavioral consequences of the interaction between EE allocation and nAChRs administration have been reported. The enriched environment diminishes the sensitivity to nicotine-induced motor effects (Adams et al., 2013; Coolon and Cain, 2009; Green et al., 2003), influences the emotional response of animals (Mesa-Gresa et al., 2013a; 2014) and decreases the sensitivity to the discriminatory stimulus characteristics of nicotine (Stairs et al., 2009). More recently, it has been demonstrated that enriched rats seem to be more sensitive to nicotine and other drugs (cocaine or amphetamine) to Conditioned Place Preference (Stairs et al., 2016). These results reflect the need of taking into account social influences and environmental conditions when evaluating effects of nicotinic agonists in preclinical models (Piña et al., 2018; Stairs et al., 2016).

This evidence also indicates the need of more in-depth studies providing a detailed perspective regarding the possible effects of EE in interaction with cholinergic agonists. In a recent study carried out by Sikora et al., (2018) the benefits of non-pharmacological methods against drug addiction have been stablished. EE strategies have been compared with interventions applied with humans and related to interventions based on physical activity, music therapy or mindfulness. These interventions have shown beneficial effects at cognitive and emotional levels, and they could be related with prevention and intervention strategies against drug abuse (Sikora et al., 2018). Following this translational point of view, future research should carried out in order to apply these beneficial effects of EE (or an active and stimulant lifestyle in humans) in combination with adequate pharmacological interventions based on alpha 7 nAChRs agonists or other pharmacotherapies for the treatment of age-related cognitive decline, neurodegenerative disorders or mental psychopathologies. In an interesting paper, reported by Patricia A. de la Tremablaye and colleagues (2018) at the University of Pittsburgh, the authors evaluated the effects of EE in combination with different pharmacotherapies in a preclinical model of traumatic brain injury. Results obtained suggested that EE induced changes both at behavioral and histological levels, but the effects of this non-pharmacological intervention were potentiated when it was combined with some pharmacotherapies such as 5-HT_{1A} agonists (de la Tremblaye et al., 2018).

Very few studies have compared the effects of nicotine or other nicotinic agonists in rodents reared in enriched or impoverished conditions. A new approach combining environmental manipulation (reflected in enriched or more impoverished environments) with drugs with specific action on alpha 7 nAChRs, such as PNU292887 or encenicline, could contribute to evaluate more in depth the role of these receptors in different behaviors as learning, memory, exploratory behavior or emotional response (Mesa-Gresa et al., 2014). Furthermore, there are several alpha 7 nAChRs agonists in different phases of clinical development (Ceskova and Silhan, 2018). Future studies with these new drugs, in combination with biomarkers and environmental influences, could also aid in the identification of better treatments.

Discussion and Conclusion

There is no effective treatment for Alzheimer's disease and the prevalence of this neurodegenerative disease is increasing. It would be necessary to develop more specific treatments aimed to improve cognitive symptoms in schizophrenia, Alzheimer's disease and in other mental and neurodegenerative disorders (Hsu et al., 2018). Experimental results obtained in different laboratories when combining a pharmacological method (administration of nicotine or nicotinic agonists) with a behavioral one (environmental manipulation as EE) could be of interest in the search and identification of new therapeutic approaches aimed to counteract or delay cognitive deficits observed in these diseases. These studies may also have translational research interest in order to obtain more information about changes in cognition and neural plasticity related to aging and neurodegenerative disorders in human subjects.

We have reviewed the main questions addressed in prior studies evaluating effects of nicotine or other nicotinic agonists on cognitive symptoms in schizophrenia, Alzheimer's disease or Parkinson's disease. Although some alpha7 nAChRs agonists have been developed and tested both in preclinical studies and in clinical trials, more research is needed in order that some of these new drugs arrived into the market (Verma et al., 2018). Studies reviewed here underscored the need of longer and controlled trials (Hsu et al., 2018). Some drugs evaluated until now have demonstrated cognitive benefits (such as encenicline for schizophrenia), but the side effects limit its possible clinical usefulness (Hsu et al., 2018). Recently, the need for better approaches based on animal models for Alzheimer`s disease has been underscored (Götz et al., 2018). These authors emphasize the need of developing a more standardized approach such as those proposed by the International Mouse Phenotyping Consortium (Götz et al., 2018).

A recent review carried out by Sale (2018) has analyzed main results regarding the key components in the application of EE to the human context. In this interesting paper, different intervention strategies related to the age of the subjects have been established. The interventions proposed include: massages for babies, early education based on cognitive stimulation and physical activity during scholar years and the maintenance of an active lifestyle combined with an enriched diet, active social relationships, physical exercise and cognitive stimulation during adulthood and for elderly persons (Sale, 2018). Different studies have obtained important results in line with the preventive and protecting effects of these enriched environments and their positive effects during life and specially, in the prevention of cognitive decline related with age and the development of neurodegenerative disorders (Cassarino and Setti, 2015; Leon and Woo, 2018; Miller et al., 2018; Mora, 2013; Sampedro-Piquero and Begega, 2017).

We consider that the approach presented in the current review could contribute in the advance of the search of therapeutic interventions for Alzheimer's or Parkinson's disease, schizophrenia and other neurodegenerative and mental disorders.

ACKNOWLEDGMENTS

This work was supported by grants from "Ministerio de Economía y Competitividad (MINECO)" (PSI2009-10410), "Conselleria d'Educació i Ciència" from Generalitat Valenciana (Spain) (PROMETEOII/2015/020) and University of Valencia (UV-INV-AE15-350056).

REFERENCES

Adams, E., Klug, J., Quast, M. and Stairs, D. J. (2013). Effects of environmental enrichment on nicotine-induced sensitization and cross-sensitization to d-amphetamine in rats. *Drug and Alcohol Dependence*, 129 (3):247-53.

Alwis, D. S. and Rajan, R. (2014). Environmental enrichment and the sensory brain: the role of enrichment in remediating brain injury. *Frontiers in System Neuroscience*, 8:156.

Bator, E., Latusz, J., Wędzony, K. and Maćkowiak, M. (2018). Adolescent environmental enrichment prevents the emergence of schizophrenia-like abnormalities in a neurodevelopmental model of schizophrenia. *European Neuropsychopharmacology*, 28 (1):97-108.

Bertrand D. and Terry A. V. Jr. (2018). The wonderland of neuronal nicotinic acetylcholine receptors. *Biochemical Pharmacology,* 151: 214-225.

Bockman, C. S., Zeng, W., Hall, J., Mittelstet, B., Schwarzkopf, L. and Stairs, D. J. (2018). Nicotine drug discrimination and nicotinic acetylcholine receptors in differentially reared rats. *Psychopharmacology*, 235 (5):1415-1426.

Burrows, E. L. and Hannan, A. J. (2016). Cognitive endophenotypes, gene-environment interactions and experience-dependent plasticity in animal models of schizophrenia. *Biological Psychology*, 116:82-89.

Carrasco, C., Vicens, P. and Redolat, R. (2006). Neuroprotective effects of behavioural training and nicotine on age-related deficits in spatial learning. *Behavioural Pharmacology*, 17 (5-6):441-52.

Cassarino, M. and Setti, A. (2015). Environment as 'Brain Training': A review of geographical and physical environmental influences on cognitive ageing. *Ageing Research Reviews*, 23 (Pt B):167-82.

Ceskova E. and Silhan P. (2018). Novel treatment options in depression and psychosis. *Neuropsychiatric Disease and Treatment*, 14:741-747.

Coolon, R. A. and Cain, M. E. (2009). Effects of mecamylamine on nicotine-induced conditioned hyperactivity and sensitization in differentially reared rats. *Pharmacology Biochemistry Behavior*, 93 (1): 59-66.

Dandi, E., Kalamari, A., Touloumi, O., Lagoudaki, R., Nousiopoulou, E., Simeonidou, C., Spandou, E. and Tata, D. A. (2018). Beneficial effects of environmental enrichment on behavior, stress reactivity and synaptophysin/BDNF expression in hippocampus following early life stress. *International Journal of Developmental Neuroscience*, 67:19-32.

de la Tremblaye, P. B., Cheng, J. P., Bondi, C. O. and Kline, A. E. (2018). Environmental enrichment, alone or in combination with various pharmacotherapies, confers marked benefits after traumatic brain injury. *Neuropharmacology*, pii: S0028-3908(18)30098-4.

Freedman, R. (2014). α7-nicotinic acetylcholine receptor agonists for cognitive enhanceent in schizophrenia. *Annual Review of Medicine*, 65:245-61.

Frozza, R. L., Lourenco, M. V. and De Felice, F. G. (2018). Challenges for Alzheimer's Disease Therapy: Insights from Novel Mechanisms Beyond Memory Defects. *Frontiers in Neuroscience,* 12:37.

Gonçalves, L. V., Herlinger, A. L., Ferreira, T. A. A., Coitinho, J. B., Pires, R. G. W. and Martins-Silva, C. (2018). Environmental enrichment cognitive neuroprotection in an experimental model of cerebral ischemia: biochemical and molecular aspects. *Behavioral Brain Research*, 348:171-183.

Götz, J., Bodea, L.G. and Goedert, M. (2018). Rodent models for Alzheimer disease. *Nature Reviews Neuroscience*, 19 (10):583-598.

Gravitz, L. (2018). Drawing on the brain's resilience to fight Alzheimer's disease. *Nature*, 559 (7715):S8-S9.

Green, T. A., Cain, M. E., Thompson, M. and Bardo, M. T. (2003). Environmental enrichment decreases nicotine-induced hyperactivity in rats. *Psychopharmacology*, 170 (3): 235-241.

Hannan, A. J. (2014). Environmental enrichment and brain repair: harnessing the therapeutic effects of cognitive stimulation and physical activity to enhance experience-dependent plasticity. *Neuropathology and Applied Neurobiology*, 40 (1):13-25.

Hsu, W. Y., Lane, H. Y. and Lin, C. H. (2018). Medications Used for Cognitive Enhancement in Patients with Schizophrenia, Bipolar Disorder, Alzheimer's Disease, and Parkinson's Disease. *Frontiers in Psychiatry*, 9:91.

Jones K. M., McDonald I. M., Bourin C., Olson R. E., Bristow L. J. and Easton A. (2014). Effect of alpha7 nicotinic acetylcholine receptor agonists on attentional set-shifting impairment in rats. *Psychopharmacology (Berl)*, 231 (4):673-83.

Jones, C. (2018). α7 Nicotinic Acetylcholine Receptor: A Potential Target in Treating Cognitive Decline in Schizophrenia. *Journal of Clinical Psychopharmacology*, (3):247-249.

Jungling, A., Reglodi, D., Karadi, Z. N., Horvath, G., Farkas, J., Gaszner, B. and Tamas, A. (2017). Effects of Postnatal Enriched Environment in a Model of Parkinson's Disease in Adult Rats. *International Journal of Molecular Sciences*, 18(2), pii: E406.

Kosik, K. S. (2015). Personalized medicine for effective Alzheimer disease treatment. *JAMA Neurology,* 72 (5):497-8.

Kucinski, A. J., Stachowiak, M. K., Wersinger, S. R., Lippiello, P. M. and Bencherif, M. (2011). Alpha7 neuronal nicotinic receptors as targets for novel therapies to treat multiple domains of schizophrenia. *Current Pharmaceutical Biotechnology, 12* (3): 437-48.

Lanctôt, K. L., Amatniek, J., Ancoli-Israel, S., Arnold, S. E., Ballard, C., Cohen-Mansfield, J., Ismail, Z., Lyketsos, C., Miller, D. S., Musiek,

E., Osorio, R. S., Rosenberg, P. B., Satlin, A., Steffens, D., Tariot, P., Bain, L. J., Carrillo, M. C., Hendrix, J. A., Jurgens, H., and Boot, B. (2017). Neuropsychiatric signs and symptoms of Alzheimer's disease: New treatment paradigms. *Alzheimers & Dementia (NY),* 3 (3):440-449.

Leon, M. and Woo, C. (2018). Environmental Enrichment and Successful Aging. *Frontiers in Behavioral Neuroscience*, 12:155.

Levin, E. D. (2012). α7-Nicotinic receptors and cognition. *Current Drug Targets*, 13 (5):602-606.

Lombardo, S. and Maskos, U. (2015). Role of the nicotinic acetylcholine receptor in Alzheimer's disease pathology and treatment. *Neuropharmacology,* 96 (PtB):255-62.

McLean, S. L., Grayson, B., Marsh, S., Zarroug, S. H., Harte, M. K. and Neill, J. C. (2016). Nicotinic α7 and α4β2 agonists enhance the formation and retrieval of recognition memory: potential mechanisms for cognitive performance enhancement in neurological and psychiatric disorders. *Behavioural Brain Research*, pii: S0166-4328(15)30167-4.

McLean, S. L., Grayson, B., Idris, N. F., Lesage, A. S., Pemberton, D. J., Mackie, C. and Neill, J. C. (2011). Activation of α7 nicotinic receptors improves phencyclidine-induced deficits in cognitive tasks in rats: implications for therapy of cognitive dysfunction in schizophrenia. *European Neuropsychopharmacology*, 21 (4):333-43.

Mesa-Gresa, P., Ramos-Campos, M. and Redolat, R. (2013a). Enriched environments for rodents and their interaction with nicotine administration. *Current Drug Abuse Reviews*, 6 (3):191-200.

Mesa-Gresa, P., Pérez-Martinez, A. and Redolat, R. (2013b). Behavioral effects of combined environmental enrichment and chronic nicotine administration in male NMRI mice. *Physiology & Behavior,* 114-115:65-76.

Mesa-Gresa, P., Pérez. Martínez, A. and Redolat, R. (2013c). Environmental enrichment improves novel object recognition and enhances agonistic behavior in male mice. *Aggressive Behaviour*, 39 (4):269-279.

Mesa-Gresa, P., Ramos-Campos, M. and Redolat, R. (2014). Behavioral effects of different enriched environments in mice treated with the cholinergic agonist PNU-282987. *Behavioural Processes*, 103:117-124.

Mesa-Gresa, P., Ramos-Campos, M. and Redolat, R. (2016). Corticosterone levels and behavioral changes induced by simultaneous exposure to chronic social stress and enriched environments in NMRI male mice. *Physiology & Behavior*, 158:6-17.

Miller, J. B., Cummings, J., Nance, C. and Ritter, A. (2018). Neuroscience learning from longitudinal cohort studies of Alzheimer's disease: Lessons for disease-modifying drug programs and an introduction to the Center for Neurodegeneration and Translational Neuroscience. *Alzheimers & Dementia* (N Y), 11 (4):350-356.

Mo, C., Hannan, A. J. and Renoir, T. (2015). Environmental factors as modulators of neurodegeneration: insights from gene-environment interactions in Huntington's disease. *Neuroscience & Biobehavioral Reviews*, 52:178-192.

Mo, C., Renoir, T. and Hannan, A. J. (2016). What's wrong with my mouse cage? Methodological considerations for modeling lifestyle factors and gene-environment interactions in mice. *Journal of Neuroscience Methods*, 265:99-108.

Mora, F. (2013). Successful brain aging: plasticity, environmental enrichment, and lifestyle. *Dialogues in Clinical Neuroscience*, 15 (1):45-52.

Nikiforuk A. (2018). Assessment of cognitive functions in animal models of schizophrenia. *Pharmacological Reports,* 70 (4):639-649.

Nithianantharajah, J. and Hannan, A. J. (2011). Mechanisms mediating brain and cognitive reserve: experience-dependent neuroprotection and functional compensation in animal models of neurodegenerative diseases. *Progress in Neuropsychopharmacology & Biological Psychiatry*, 35 (2):331-339.

Novaes, L. S., Dos Santos, N. B., Batalhote, R. F. P., Malta, M. B., Camarini, R., Scavone, C. and Munhoz, C. D. (2017). Environmental enrichment protects against stress-induced anxiety: Role of glucocorticoid receptor, ERK, and CREB signaling in the basolateral amygdala. *Neuropharmacology*, 113 (Pt A):457-466.

Pang, T. Y. and Hannan, A. J. (2013). Enhancement of cognitive function in models of brain disease through environmental enrichment and physical activity. *Neuropharmacology*, 64:515-28.

Pfeffer A., Munder T., Schreyer S., Klein C., Rasińska J., Winter Y. and Steiner B. (2018). Behavioral and psychological symptoms of dementia (BPSD) and impaired cognition reflect unsuccessful neuronal compensation in the pre-plaque stage and serve as early markers for Alzheimer's disease in the APP23 mouse model. *Behavioural Brain Research,* 16 (347):300-313.

Piña, J. A., Namba, M. D., Leyrer-Jackson, J. M., Cabrera-Brown, G. and Gipson, C.D. (2018). Social Influences on Nicotine-Related Behaviors. *International Review of Neurobiology*, 140:1-32.

Prado Lima, M. G., Schimidt, H. L., Garcia, A., Daré, L. R., Carpes, F. P., Izquierdo, I. and Mello-Carpes, P. B. (2018). Environmental enrichment and exercise are better than social enrichment to reduce memory deficits in amyloid beta neurotoxicity. *Proceedings of the National Academy of Sciences of USA*, 115 (10):E2403-E2409.

Redolat, R. and Mesa-Gresa, P. (2012). Potential benefits and limitations of enriched environments and cognitive activity on age-related behavioural decline. *Current Topics in Behavioral Neuroscience*, 10:293-316.

Rogers, J., Renoir, T. and Hannan, A. J. (2017). Gene-environment interactions informing therapeutic approaches to cognitive and affective disorders. *Neuropharmacology,* pii: S0028-3908(17)30636-6.

Sale, A. (2018). A Systematic Look at Environmental Modulation and Its Impact in Brain Development. *Trends in Neurosciences*, 41 (1):4-17.

Sale, A., Berardi, N. and Maffei, L. (2014). Environment and brain plasticity: towards an endogenous pharmacotherapy. *Physiological Reviews*, 94 (1):189-234.

Sampedro-Piquero, P. and Begega, A. (2017). Environmental Enrichment as a Positive Behavioral Intervention Across the Lifespan. *Current Neuropharmacology*, 15 (4):459-470.

Seong, H. H., Park, J. M. and Kim, Y. J. (2018). Antidepressive Effects of Environmental Enrichment in Chronic Stress-Induced Depression in Rats. *Biological Research for Nursing*, 20 (1):40-48.

Sikora, M., Nicolas, C., Istin, M., Jaafari, N., Thiriet, N. and Solinas, M. (2018). Generalization of effects of environmental enrichment on seeking for different classes of drugs of abuse. *Behavioral Brain Research*, 341:109-113.

Skwara, A. J., Karwoski, T. E., Czambel, R. K., Rubin, R. T. and Rhodes, M. E. (2012). Influence of environmental enrichment on hypothalamic-pituitary-adrenal (HPA) responses to single-dose nicotine, continuous nicotine by osmotic mini-pumps, and nicotine withdrawal by mecamylamine in male and female rats. *Behavioural Brain Research*, 234 (1): 1-10.

Stairs, D. J. and Bardo, M. T. (2009). Neurobehavioral effects of environmental enrichment and drug abuse vulnerability. *Pharmacology, Biochemistry and Behavior*, 92 (3):377.

Stairs, D. J., Kangiser, M., Hickle, T. and Bockman, C. S. (2016). Effects of Environmental Enrichment on Nicotine Addiction. In *Foundations of Understanding, Tobacco, Alcohol, Cannabinoids and Opioids* (Vol. 1, pp. 246-253). Elsevier Inc.

Titova, N. and Chaudhuri. K. R. (2018). Non-motor Parkinson disease: new concepts and personalised management. *Medical Journal of Australia*, 208 (9):404-409.

Tregellas, J. R. and Wylie, K. P. (2018). Alpha7 Nicotinic Receptors as Therapeutic Targets in Schizophrenia. *Nicotine & Tobacco Research.* In press.

Verma, S., Kumar, A., Tripathi, T. and Kumar, A. (2018). Muscarinic and nicotinic acetylcholine receptor agonists: current scenario in Alzheimer's disease therapy. *Journal of Pharmacy & Pharmacology*, 70 (8):985-993.

Yang, T., Xiao, T., Sun, Q. and Wang, K. (2017). The current agonists and positive allosteric modulators of α7 nAChR for CNS indications in clinical trials. *Acta Pharmaceutica Sinica B,* 7 (6):611-622.

Young, J. W. and Geyer, M. A. (2013). Evaluating the role of the alpha-7 nicotinic acetylcholine receptor in the pathophysiology and treatment of schizophrenia. *Biochemical Pharmacology*, 15, 86(8):1122-32.

Valentine G. and Sofuoglu M. (2018). Cognitive Effects of Nicotine: Recent Progress. *Current Neuropharmacology,* 16 (4):403-414.

Zhu, J., Bardo, M. T., Green, T. A., Wedlund, P. J. and Dwoskin, L. P. (2007). Nicotine increases dopamine clearance in medial prefrontal cortex in rats raised in an enriched environment. *Journal of Neurochem*istry, 103 (6): 2575-88.

Zhu, J., Bardo, M. T. and Dwoskin, L. P. (2013). Distinct effects of enriched environment on dopamine clearance in nucleus accumbens shell and core following systemic nicotine administration. *Synapse*, 67 (2): 57-67.

BIBLIOGRAPHY

Accounts in drug discovery: case studies in medicinal chemistry

LCCN	2012405578
Type of material	Book
Main title	Accounts in drug discovery: case studies in medicinal chemistry / edited by Joel C. Barrish ... [et al.].
Published/Created	Cambridge: Royal Society of Chemistry, c2011.
Description	xvi, 379 p.: ill. (some col.); 24 cm.
ISBN	9781849731263 (hbk.) 1849731268 (hbk.)
LC classification	RM301.25 .A23 2011
Related names	Barrish, Joel Charles, 1957- Royal Society of Chemistry (Great Britain)
Contents	The discovery of the dipeptidyl peptidase-4 (DPP4) inhibitor Onglyza: from concept to market / Jeffrey A. Robl and Lawrence G. Hamann -- The discovery of Vorapaxar (SCH 530348), a thrombin receptor (protease activated receptor-1) antagonist with potent antiplatelet effects / Samuel Chackalamannil -- The

discovery of Piragliatin, a glucokinase activator / Ramakanth Sarabu, Jefferson W. Tilley and Joseph Grimsby -- The discovery of OSI-906, a small-molecule inhibitor of the insulin-like growth factor-1 and insulin receptors / Mark J. Mulvihill and Elizabeth Buck -- The genesis of the antibody conjugate gemtuzumab ozogamicin (Mylotarg®) for acute myeloid leukemia / Philip R. Hamann -- Novel androgen receptor antagonists for the treatment of prostate cancer / Ashvinikumar V. Gavai, William R. Foster, Aaron Balog and Gregory D. Vite -- The discovery of UK-390957: the challenge of targeting a short half-life, rapid Tmax SSRI / Mark D. Andrews and Donald S. Middleton -- The discovery of the long-acting PDE5 inhibitor PF-489791 for the treatment of pulmonary hypertension / Andrew S. Bell and Michael J. Palmer -- From HTS to market: the discovery and development of maraviroc, a CCR5 antagonist for the treatment of HIV / Chris Barber and David Pryde -- The discovery of GS-9131, an amidate prodrug of a novel nucleoside phosphate HIV reverse transcriptase inhibitor / Richard Mackman -- 2'-F-2'-C-methyl nucleosides and nucleotides for the treatment of hepatitis C virus: from discovery to the clinic / Michael J. Sofia, Phillip A. furman and William T. Symonds -- A case history on the challenges of central nervous system and dual pharmacology drug discovery / Paul V. Fish, Anthony Harrison, Florian Wakenhut and Gavin A. Whitlock -- The discovery of TRPV1 antagonists: turning up the heat / Mark H.

	Norman -- Uncoupling neuroprotection from immunosuppression: the discovery of ILS-920 / Edmund I. Graziani -- Identification of [alpha] 7 nicotinic acetylcholine receptor agonists for their assessment in improving cognition in schizophrenia / Bruce N. Rogers ... [et al.]
Subjects	Drug development--Case studies.
Notes	Includes bibliographical references and index.
Series	RSC drug discovery series; no. 4.

Cholesterol regulation of ion channels and receptors

LCCN	2012011095
Type of material	Book
Main title	Cholesterol regulation of ion channels and receptors / edited by Irena Levitan, Francisco Barrantes.
Published/Created	Hoboken, N.J.: John Wiley & Sons, c2012.
Description	xii, 289 p., [20] p.: ill. (some col.); 24 cm.
ISBN	9780470874325 (cloth) 0470874325 (cloth)
LC classification	QP752.C5 C46 2012
Related names	Levitan, Irena. Barrantes, Francisco J., 1944-
Contents	Cholesterol trafficking and distribution between cellular membranes / Daniel Wustner, Lukasz M. Solanko, Frederik W. Lund -- Cholesterol regulation of membrane protein function by changes in bilayer physical properties: an energetic perspective / Jens A. Lundbaek and Olaf S. Andersen -- Insights into structural determinants of cholesterol sensitivity of Kir channels / Avia Rosenhouse-Dantsker and Irena Levitan -- Role for lipid rafts in the regulation of store-operated Ca2+ channels / Hwei L. Ong and

	Indu S. Ambudkar -- Cholesterol regulation of cardiac ion channels / Elise Balse, Stephane Hatem and Stanley Nattel -- Differential contribution of BK subunits to nongenomic regulation of channel function by steroids / Alex M. Dopico, Anna N. Bukiya, and Aditya K. Singh -- Regulation of K+ channels by cholesterol-rich membrane domains in immune system / Núria Comes and Antonio Felipe -- Indirect channel regulation by cholesterol: the role of caveolae and caveolins in regulating KATP channel function / Caroline Dart -- Regulation of the nicotinic acetylcholine receptor by cholesterol as a boundary lipid / Francisco J. Barrantes -- Specific and non-specific regulation of GPCR function by cholesterol / Gerald Gimpl and Katja Gehrig-Burger -- Structural insights into cholesterol interactions with G protein-coupled receptors / Jeremiah S. Joseph, Enrique E. Abola, and Vadim Cherezov -- Membrane cholesterol: implications in receptor function / Sandeep Shrivastava and Amitabha Chattopadhyay -- The role of cholesterol and lipid rafts in regulation of TLR receptors / Ruxana T. Sadikot.
Subjects	Cholesterol--Metabolism. Ion channels--Metabolism. Membrane proteins--Metabolism.
Notes	Includes bibliographical references and index.

Dopamine-glutamate interactions in the basal ganglia

LCCN	2011039590
Type of material	Book
Main title	Dopamine-glutamate interactions in the basal

	ganglia / edited by Susan Jones.
Published/Created	Boca Raton, FL: CRC Press, c2012.
Description	xxiii, 260 p.: ill. (some col.); 24 cm.
ISBN	9781420088793 (hardback: alk. paper) 1420088793 (hardback: alk. paper)
LC classification	QP383.3 .D67 2012
Related names	Jones, Susan, Ph. D.
Summary	"An exploration of the nature of dopamine-glutamate interactions in the basal ganglia from receptor molecules to complex behaviors, this volume reviews basic anatomy, discusses the subtypes and signaling pathways of the dopamine and glutamate receptors expressed in the basal ganglia, and their interaction down to the molecular level. Coverage includes endogenous and exogenous modulators of dopamine-glutamate interactions and the implications of these interactions, measuring the key basal ganglia functions at the physiological and behavioral level. It also examines the concept that compromised dopamine-glutamate interactions may underpin basal ganglia disorders"-- Provided by publisher.
Contents	Ch. 1. Metabotropic glutamate receptor-dopamine interactions in the basal ganglia motor circuit / Kari A. Johnson and P. Jeffrey Conn -- ch. 2. Ionotropic glutamate receptors in the basal ganglia / Susan Jones [and others] -- ch. 3. Dopamine receptors and their interactions with NMDA receptors / Alasdair J. Gibb and Huaxia Tong -- ch. 4. Synaptic triad in the neostriatum: dopamine, glutamate, and the MSN / Jason R. Klug [and others] -- ch. 5. Dopaminergic modulation of glutamatergic synaptic plasticity

	in striatal circuits: new insights from BAC-transgenic mice / D. James Surmeier [and others] -- ch. 6. Striatal acetylcholine-dopamine crosstalk and the dorsal-ventral divide / Sarah Threlfell and Stephanie J. Cragg -- ch. 7. Electrophysiology of the corticostriatal network in Vivo / Morgane Pidoux, Sãaeverine Mahon, and Stéphane Charpier -- ch. 8. Functional organization of the midbrain substantia nigra / Jennifer Brown -- ch. 9. Striatal dopamine and glutamate in action: the generation and modification of adaptive behavior / Henry H. Yin and Rui M. Costa -- ch. 10. Impaired dopamine-glutamate receptor interactions in some neurological disorders / Miriam A. Hickey and Carlos Cepeda.
Subjects	Basal ganglia--Physiology. Basal ganglia--Pathophysiology. Dopamine. Neurotransmitters. Basal Ganglia--physiology. Basal Ganglia Diseases--physiopathology. Dopamine--physiology. Receptors, Glutamate.
Notes	Includes bibliographical references and index.
Series	Frontiers in neuroscience

FASEB bioAdvances.

LCCN	2017202855
Type of material	Periodical or Newspaper
Main title	FASEB bioAdvances.
Published/Produced	Hoboken, NJ: John Wiley & Sons, Inc., 2018-
Current frequency	Twelve issues a year
ISSN	2573-9832

LC classification	QH324
Serial key title	FASEB bioAdvances
Abbreviated title	FASEB bioAdvances
Related names	Federation of American Societies for Experimental Biology.
Subjects	Biology, Experimental--Periodicals.
Form/Genre	Electronic journal.
Notes	Some articles may be unedited but they are citable. The final edited and typset version of record will appear in the future. The article has been accepted for publication and undergone full peer review but has not been through the copyediting, typesetting, pagination and proofreading process, which may lead to differences between unedited version and the Version of Record.

Imaging of the human brain in health and disease

LCCN	2013039184
Type of material	Book
Main title	Imaging of the human brain in health and disease / edited by Philip Seeman, Bertha Madras.
Edition	First edition.
Published/Produced	Amsterdam; Boston: Academic Press/Elsevier, 2014.
Description	xiv, 517 pages: illustrations (some color); 25 cm.
ISBN	9780124186774 (alk. paper) 0124186777 (alk. paper)
LC classification	RC386.6.T65 I43 2014
Related names	Seeman, Philip, editor of compilation. Madras, Bertha, editor of compilation.
Summary	"Modern imaging techniques have allowed researchers to non-invasively peer into the human brain and investigate, among many other

things, the acute effects and long-term consequences of drug abuse. Here, we review the most commonly used and some emerging imaging techniques in addiction research, explain how the various techniques generate their characteristic images and describe the rational that researchers use to interpret them. In addition, examples of seminal imaging findings are highlighted that illustrate the contribution of each imaging modality to the expansion in our understanding of the neurobiological bases of drug abuse and addiction, and how they can be parlayed in the future into clinical and therapeutic applications"-- Provided by publisher.

Contents

Machine generated contents note: 1. Neuroimaging of Addiction 2. Brain Imaging of Sigma Receptors 3. Imaging of Neurochemical Transmission in the Central Nervous System 4. Human Brain Imaging of Acetylcholine Receptors 5. Human Brain Imaging of Opioid Receptors: Application to CNS Biomarker and Drug Development 6. Human Brain Imaging of Adenosine Receptors 7. Human Brain Imaging of Dopamine D1 Receptors 8. Human Brain Imaging of Dopamine Transporters 9. Dopamine Receptor Imaging in Schizophrenia: Focus on Genetic Vulnerability 10. Human Brain Imaging of Anger 11. Imaging Pain in the Human Brain 12. Imaging of Dopamine and Serotonin Receptors and Transporters 13. Imaging the Dopamine D3 Receptor In Vivo 14. Human Brain Imaging of Autism Spectrum Disorders 15. Brain PET Imaging in the Cannabinoid System

	16. Brain Imaging of Cannabinoid Receptors 17. Human Brain Imaging In Tardive Dyskinesia 18. Dopamine Receptors and Dopamine Release 19. Radiotracers Used to Image the Brains of Patients with Alzheimer's Disease.
Subjects	Brain--Tomography. Brain--Diseases--Diagnosis. Neuroimaging--methods. Brain Chemistry--physiology. Brain Diseases--radionuclide imaging. Mental Disorders--radionuclide imaging. Substance-Related Disorders--radionuclide imaging.
Notes	Includes bibliographical references and index.
Series	Neuroscience-net reference book series; book 1

Insect nicotinic acetylcholine receptors

LCCN	2010011799
Type of material	Book
Main title	Insect nicotinic acetylcholine receptors / edited by Steeve Hervé Thany.
Published/Created	New York: Springer Science+Business Media; Austin, Tex.: Landes Bioscience, c2010.
Description	xix, 118 p.: ill.; 26 cm.
Links	Contributor biographical information http://www.loc.gov/catdir/enhancements/fy1106/2010011799-b.html Publisher description http://www.loc.gov/catdir/enhancements/fy1106/2010011799-d.html Table of contents only http://www.loc.gov/catdir/enhancements/fy1106/2010011799-t.html
ISBN	9781441964441 1441964444

LC classification	QP364.7 .I565 2010
Related names	Thany, Steeve Herve, 1972-
Subjects	Nicotinic receptors. Insecticides--Physiological effect. Insects--Physiology. Receptors, Nicotinic. Insecticides--toxicity. Synaptic Transmission.
Notes	Includes bibliographical references and index.
Series	Advances in experimental medicine and biology; v. 683

Muscarinic receptor: from structure to animal models

LCCN	2015945343
Type of material	Book
Main title	Muscarinic receptor: from structure to animal models / edited by Jaromir Myslivecek, Institute of Physiology, Charles University, Prague, Czech republic, Jan Jakubik, Department of Neurochemistry, Achedemy of Schiences of the Czech Republic, Institute of Physiology, Prague, Czech Republic.
Published/Produced	New York, NY: Humana Press, 2016 ©2016
Description	xii, 287: illustrations (some color); 26 cm
ISBN	9781493928576 (hbk) 1493928570 (hbk) ebook 1493928589
LC classification	QP364.7 .M848 2016
Related names	Myslivecek, Jaromir, editor of compilation. Jakubík, Ján, 1923- editor of compilation.
Contents	Machine generated contents note: 1.Towards the Crystal Structure Determination of Muscarinic

	Acetylcholine Receptors / Takuya Kobayashi -- 2.Homology Modeling and Docking Evaluation of Human Muscarinic Acetylcholine Receptors / Elizabeth Yuriev -- 3.Radioligand Binding at Muscarinic Receptors / Jan Jakubik -- 4.Binding Method for Detection of Muscarinic Acetylcholine Receptors in Receptor's Natural Environment / Matomo Nishio -- 5.Use of Antibodies in the Research on Muscarinic Receptor Subtypes / Martin C. Michel -- 6.Allosteric Modulation of Muscarinic Receptors / Esam E. El-Fakahany -- 7.Subcellular and Synaptic Localization of Muscarinic Receptors in Neurons Using High-Resolution Electron Microscopic Preembedding Immunogold Technique / Veronique Bernard -- 8.Investigation of Muscarinic Receptors by Fluorescent Techniques / Moritz Bunemann -- 9.Autoradiography Assessment of Muscarinic Receptors in the Central Nervous System / Jaromir Myslivecek --
Subjects	Muscarinic receptors--Laboratory manuals. Muscarinic receptors.
Form/Genre	Laboratory manuals.
Notes	Includes bibliographical references and index.
Additional formats	Also available in an electronic edition.
Series	Neuromethods, 0893-2336; 107 Springer protocols

Muscarinic receptors

LCCN	2011943086
Type of material	Book
Main title	Muscarinic receptors / Allison D. Fryer, Arthur Christopoulos, Neil M. Nathanson, editors.

Published/Created	Heidelberg: Springer, c2012.
Description	xiii, 499 p.: ill.
ISBN	9783642232732 (alk. paper) 3642232736 (alk. paper) 9783642232749 (e-ISBN) 3642232744 (e-ISBN)
LC classification	QP364.7 .M87 2012
Related names	Fryer, Allison D. Christopoulos, Arthur. Nathanson, Neil M.
Contents	Muscarinic receptor pharmacology and signaling -- Overview of muscarinic receptor subtypes -- Structure-function studies of muscarinic acetylcholine receptors -- Polymorphisms in human muscarinic receptor subtype genes -- Muscarinic receptor trafficking -- Physiological role of G-protein coupled receptor phosphorylation -- Novel muscarinic receptor mutant mouse models -- Muscarinic receptors in the CNS -- Muscarinic receptor pharmacology and circuitry for the modulation of cognition -- Muscarinic agonists and antagonists in schizophrenia -- Muscarinic pain pharmacology: realizing the promise of novel anagesics by overcoming old challenges -- Muscarinic modulation of striatal function and circuitry -- Muscarinic receptors in brain stem and mesopontine cholinergic arousal functions -- Muscarinic receptors in autonomic effector organs -- Muscarinic receptor agonists and antagonists: effects on ocular function -- Muscarinic receptor agonists and antagonists: effects on cardiovascular function -- Muscarinic receptor antagonists: effects on pulmonary

	function -- Muscarinic agonists and antagonists: effects on gastrointestinal function -- Muscarinic agonists and antagonists: effets on the urinary bladder -- Muscarinic receptors and mediation of hormonal effects of acetylcholine -- Muscarinic receptor agonists and antagonists: effects on inflammation and immunity -- Muscarinic receptor agonists and antagonists: effects on keratinocyte functions -- Muscarinic receptor agonists and antagonists: effects on cancer -- Activation of muscarinic receptors by non-neuronal acetylcholine.
Subjects	Muscarinic receptors. Receptors, Muscarinic.
Notes	Includes bibliographical references and index.
Series	Handbook of experimental pharmacology, 0171-2004; v. 208

Nicotinic acetylcholine receptor technologies

LCCN	2016943316
Type of material	Book
Main title	Nicotinic acetylcholine receptor technologies / edited by Ming D. Li, University of Virginia, Charlottesville, VA, USA, Zhejiang University of Hangzhou, Zhejing, China.
Published/Produced	New York: Humana Press, [2016]
Description	xii, 259 pages: illustrations; 27 cm
Links	http://www.springerprotocols.com/BookToc/doi/10.1007/978-1-4939-3768-4
ISBN	9781493937660 (hbk.: alk. paper) 1493937669 (hbk.: alk. paper)
LC classification	QP364.7 .N53 2016
Related names	Li, Ming D., editor.
Summary	This volume provides properties, biological

function, methods, and approaches for manipulating nAChRs in different organisms. Nicotinic Acetylcholine Receptor Technologies guides readers through molecular techniques and behavioral tests used to investigate nicotinic drugs, chronoamperometry, emerging technologies and methods for the analysis of nAChRs, single nucleotide polymorphism (SNP), fluorescence techniques, spectral confocal, allosteric modulators of a7-nAChRs, and a comprehensive evolutional relation for most nAChR subunits, in both vertebrate and invertebrate species. Written for the popular Neuromethods series, chapters include the kind of detail and key implementation advice that ensures successful results in the laboratory. Authoritative and practical, Nicotinic Acetylcholine Receptor Technologies aims to ensure successful results in the further study of this vital field.-- Source other than Library of Congress.

Subjects	Nicotinic receptors.
	Acetylcholine--Receptors.
	Receptors, Nicotinic--genetics.
	Mental Disorders--drug therapy.
	Mental Disorders--genetics.
	Molecular Targeted Therapy.
	Models, Animal.
	Acetylcholine--Receptors.
	Nicotinic receptors.
Form/Genre	Laboratory Manuals.
Notes	Includes bibliographical references and index.
Series	Neuromethods, 0893-2336; 117
	Springer protocols

Nicotinic receptors

LCCN	2014947538
Type of material	Book
Main title	Nicotinic receptors / Robin A.J. Lester, editor.
Published/Produced	New York: Humana Press, [2014]
Description	xvii, 461 pages: illustrations (some color); 24 cm.
ISBN	9781493911660 (alk. paper) 149391166X (alk. paper)
LC classification	QP364.7 .N543 2014
Related names	Lester, Robin A. J., editor.
Summary	A comprehensive overview of nicotinic receptors that addresses their history from crystal structure to behavior as well as their implications in disease and potential as therapeutic targets. It includes background information on all subtypes of nicotinic receptors, the most recent information on the distribution throughout the nervous system, and discussion of their implications in learning and memory, addiction, and neurological and psychiatric disease such as Alzheimer's and Parkinson's. Takes advantage of several recent developments in the fields of optogenetics, viral expression, and gene analysis to focus on current knowledge on the functional aspects of nicotinic receptors. -- Source other than Library of Congress.
Contents	On the discovery of the nicotinic acetylcholine receptor channel -- Molecular structure, gating, and regulation -- Molecular underpinnings of neuronal nicotinic acetylcholine receptor expression -- Presynaptic nicotinic acetylcholine receptors: subtypes and functions -- Functional distribution and regulation of neuronal nicotinic

	ACh receptors in the mammalian brain -- Nicotinic signaling in development -- Presynaptic nicotinic acetylcholine receptors and the modulation of circuit excitability -- Autonomic nervous system transmission -- Nicotinic receptors in the spinal cord -- Slow synaptic transmission in the central nervous system -- The effects of nicotine on learning and memory -- Nicotinic receptors as targets for novel analgesics and anti-inflammatory drugs -- Nicotinic acetylcholine receptors and the roles of the Alpha7 subunit -- Role of central serotonin receptors in nicotine addiction -- Neuronal nicotinic acetylcholine receptors in reward and addiction -- Genetic contributions of the a5 nicotinic receptor subunit to smoking behavior -- Smoking-related genes and functional consequences -- Nicotinic acetylcholine receptors along the habenulo-interpeduncular pathway: roles in nicotine withdrawal and other aversive aspects -- Nicotinic acetylcholine receptors in Alzheimer's and Parkinson's disease -- Nicotinic receptors and mental illness -- Current and future trends in drug discovery and development related to nicotinic receptors
Subjects	Nicotinic receptors.
Notes	Includes bibliographical references.
Series	The receptors; volume 26

Targets and emerging therapies for schizophrenia

LCCN	2011051091
Type of material	Book
Main title	Targets and emerging therapies for schizophrenia / edited by Jeffrey S. Albert, Michael W. Wood.

Published/Created	Hoboken, N.J.: Wiley, c2012.
Description	xiii, 472 p.: ill. (some col.); 24 cm.
ISBN	9780470322826 (cloth)
	0470322829 (cloth)
LC classification	RC514 .T37 2012
Related names	Albert, Jeffrey S.
	Wood, Michael W. (Michael William), 1961-
Contents	Dopaminergic hypothesis of schizophrenia: a historical perspective / Aurelija Jucaite and Svante Nyberg -- Dopamine D2/D3 partial agonists as antipsychotics / Philip G. Strange -- D1/D5 dopamine agonists as pharmacotherapy for schizophrenia / Kevin N. Boyd and Richard B. Mailman -- PDE inhibitors as a novel therapeutic approach for schizophrenia / Judith A. Siuciak and William J. Pitts -- Glutamatergic synaptic dysregulation in schizophrenia / Joseph T. Coyle, Alo Basu, and Michael Benneyworth -- Metabotropic glutamate 2/3 receptor agonists and positive allosteric modulators of metabotropic glutamate receptor 2 as novel agents for the treatment of schizophrenia / Gerard J. Marek ... [et al.] -- AMPA receptor positive modulators / John A. Morrow, John K.F. Maclean, and Craig Jamieson -- Progress in the exploration and development of GlyT1 inhibitors for schizophrenia / Jeffrey S. Albert and Michael W. Wood -- Combined dopamine D2 and 5-hydroxytryptamine (5-HT)1a receptor strategies for the treatment of schizophrenia, a pharmacological and chemical perspective / Andrew C. Mccreary, Roelof W. Feenstra, and Caitlin A. Jones -- 5-HT2C and 5-HT6 receptor targeted emerging approaches in schizophrenia /

	Sharon Rosenzweig-Lipson ... [et al.] -- The cholinergic hypothesis: an introduction to the hypothesis and a short history / Joseph I. Friedman, Isabella Kanellopoulou, and Vladan Novakovic -- À7 nicotinic acetylcholine receptors in the treatment of schizophrenia / Mihály Hajós, and Bruce N. Rogers -- Muscarinic acetylcholine receptors as novel targets for the development of therapeutics for schizophrenia / Christian C. Felder ... [et al.] -- Will modulation of neuropeptide receptors produce the next generation of antipsychotic drugs?: a focus on the neurokinin and neurotensin systems / Lee A. Dawson, Paul W. Smith, and Jeannette M. Watson -- GABA and schizophrenia / John H. Kehne and George D. Maynard.
Subjects	Schizophrenia--Chemotherapy. Drug targeting. Schizophrenia--drug therapy.
Notes	Includes bibliographical references and index.

Textbook of drug design and discovery

LCCN	2009013698
Type of material	Book
Main title	Textbook of drug design and discovery / edited by Povl Krogsgaard-Larsen, Kristian Strømgaard, Ulf Madsen.
Edition	4th ed.
Published/Created	Boca Raton, FL: CRC Press/Taylor & Francis, c2010.
Description	xv, 460 p.: ill. (some col.); 27 cm.
ISBN	9781420063226 (alk. paper) 1420063227 (alk. paper)

9781420063264 (alk. paper)
142006326X (alk. paper)

LC classification RS420 .T49 2010

Related names Krogsgaard-Larsen, Povl.
Strømgaard, Kristian.
Madsen, Ulf.

Contents Molecular recognition in ligand-protein binding / Tommy Liljefors -- Biostructure-based drug design / Flemming S. Jørgensen and Jette S. Kastrup -- Ligand-based drug design / Ingrid Pettersson, Thomas Balle, and Tommy Liljefors -- Chemical biology / Kristian Strømgaard -- Stereochemistry in drug design / M. Belen Mayo-Martin and David E. Jane -- Natural products in drug discovery / Guy T. Carter -- Imaging in drug discovery and development / Markus Rudin and Thomas Mueggler -- Peptides and peptidomimetics / Minying Cai, Vinod Kulkarni, and Victor J. Hruby -- Prodrugs: design and development / Anders Buur and Niels Mørk -- Metals in medicine: inorganic medicinal chemistry / Helle R. Hansen and Ole Farver -- Enzyme inhibitors: biostructure-based and mechanism-based designs / Robert A. Copeland, Richard R. Gontarek, and Lusong Luo -- Receptors: structure, function, and pharmacology / Hans Bräuner-Osborne -- Ion channels: structure and function / Søren-Peter Olesen and Daniel B. Timmermann -- Neurotransmitter transporters: structure and function / Claus J. Loland and Ulrik Gether -- GABA and glutamic acid receptor ligands / Bente Frølund and Ulf Madsen -- Acetylcholine / Anders A. Jensen and Povl Krogsgaard-Larsen -- Histamine receptors /

	Iwan de Esch, Henk Timmerman and Rob Leurs -- Dopamine and serotonin / Benny Bang-Andersen and Klaus P. Bøgesø -- Opioid and cannabinoid receptors / Rasmus P. Clausen and Harald S. Hansen -- Hypnotics / Bjarke Ebert and Keith Wafford -- Neglected diseases / Søren B. Christensen -- Immunomodulating agents / Ulla G. Sidelmann -- Anticancer agents / Fredrik Bjørkling and Lars H. Jensen -- Antiviral drugs / Erik De Clercq -- Antibiotics / Piet Herdewijn.
Subjects	Drugs--Design. Drug Discovery.
Notes	Includes bibliographical references and index.

The birthday problem

LCCN	2014902250
Type of material	Book
Personal name	Gussoff, Caren, author.
Main title	The birthday problem / Caren Gussoff.
Edition	First trade paperback edition.
Published/Produced	Auburn, MA: Pink Narcissis Press, 2014.
Description	216 pages; 22 cm
ISBN	9781939056061 (pbk.) 1939056063 (pbk.)
LC classification	PS3557.U8336 B57 2014
Summary	"In the year 2060, the next plague has arrived. MaGo bots, the nanotechnology used for everything from fighting the common cold to radical life extension, have begun to malfunction, latching onto the brain's acetylcholine receptors to cause a permanent state of delirium. The Birthday Problem follows four Seattle survivors: Chaaya Gopal Lee, great-granddaughter of the MaGo programmer, whom

the pandemic turns into a killer; 40-something ex-rock star and pharmacy technician Greystone Toussaint, the 'King of Seattle'; Alastair Gomez-Larsen, forced to become a blood-smuggler to treat his father's liver disease; and Didi VanNess, a lovesick former-WNBA center and CNA, who tries to win back her wife's heart against a backdrop of madness, death and 30 cats, all named Ira."--Back cover.

Subjects	Nanotechnology--Fiction.
	Epidemics--Fiction.
	Dystopias--Fiction.
	Seattle (Wash.)--Fiction.
Form/Genre	Science fiction.

The cardiovascular system: morphology, control and function

LCCN	2018285366
Type of material	Book
Main title	The cardiovascular system: morphology, control and function / [edited by] A. Kurt Gamperl, Departments of Ocean Sciences and Biology, Memorial University of Newfoundland, St. John's, Newfoundland and Labrador, Canada, Todd E. Gillis, Department of Integrative Biology, University of Guelph, Guelph, Ontario, Canada, Anthony P. Farrell, Department of Zoology, and Faculty of Land and Food Systems, The University of British Columbia, Vancouver, British Columbia, Canada, Colin J. Brauner, Department of Zoology, The University of British Columbia, Vancouver, British Columbia, Canada.
Edition	First edition
Published/Produced	Cambridge, MA: Elsevier, Academic Press,

	2017.
	©2017
Description	xxv, 452 pages: illustrations (some color), charts; 24 cm.
ISBN	9780128041635 (pbk.) 0128041633 (pbk.)
LC classification	QL639.1 .C37 2017 QL639.1 .F58 v.36A
Related titles	The cardiovascular system. Development, plasticity and physiological responses. Complemented by (work):
Related names	Gamperl, A. Kurt (Anthony Kurt), editor. Gillis, Todd E., editor. Farrell, Anthony Peter, 1952- contributor, editor. Brauner, Colin J., editor.
Summary	"Volume 36A ... summarizes our current understanding of the fish heart and vasculature, and how they work"--Back cover.
Contents	Contributors -- Abbreviations -- Preface -- Heart morphology and anatomy / José M. Icardo -- 1. Introduction -- 2. Fish heart chambers: a reassessment -- 3. Sequential analysis of the heart: a comparative approach -- 3.1. The sinus venosus -- 3.2. The atrium -- 3.3. The AV segment -- 3.4. The ventricle -- 3.5. The outflow tract -- 3.5.1. Basal gnathostomata -- 3.5.2. More advanced teleosts -- 3.5.2.1. The conus arteriosus and the conus calves -- 3.5.2.2. The bulbus arteriosus -- 4. Blood supply to the heart chambers -- 5. Cardiac nerves -- 6. The heart's pacemaker and conduction system -- 7. Lungfish heart: a special case -- 7.1. The sinus venosus -- 7.2. The atrium, the AV region and the ventricle -- 7.3. The outflow tract -- 8. Summary and

future directions -- Acknowledgments -- References -- Cardiomyocyte morphology and physiology / Holly A. Shiels -- 1. Introduction -- 2. Gross myocyte morphology -- 2.1. Sarcolemmal and cell-cell interactions -- 2.1.1. Mechanical connections between cells -- 2.1.2. Electrical Connections Between Cells -- 2.2. Mitochondria -- 3. Excitation-contraction coupling -- 3.1. Extracellular Ca2+ influx -- 3.1.1. The L-Type Ca2+ channel -- 4. ss-Adrenergic receptors -- 5. The myofilaments -- 6. Conclusions -- References -- Electrical excitability of the fish heart and its autonomic regulation / Matt Vornanen -- 1. Introduction -- 2. Electrical excitability of the fish heart -- 3. Cardiac action potential -- 4. Rhythm of the heartbeat and impulse conduction -- 5. Ion currents of the fish heart -- 5.1. Inward currents -- 5.1.1. Sodium current (INa) -- 5.1.2. Calcium currents (ICa) -- 5.1.2.1. L-type Ca2+ current (ICaL) -- 5.1.2.2. T-type Ca2+ current (ICaT) -- 5.1.2.3. The hyperpolarization-activated funny current (If) -- 5.2. Outward potassium currents -- 5.2.1. Voltage-gated K+ currents (IK) -- 5.2.1.1. The rapid component of the delayed rectifier (IKr) -- 5.2.1.2. The slow component of the delayed rectifier (IKs) -- 5.2.2. Inward rectifier K+ currents (IKir) -- 5.2.2.1. The background inward rectifier current (IK1) -- 5.2.2.2. Acetylcholine-activated inward rectifier current (IKACH) -- 5.2.2.3. ATP-sensitive potassium current (IKATP) -- 6. Effects of autonomic nervous control on cardiac excitability -- 6.1. Cholinergic regulation of nodal tissues and the atrium -- 6.2. Adrenergic regulation -- 7. Significance of ion channel function in thermal tolerance of fish hearts -- 7.1. Thermal tolerance limits of the fish heart in comparison with other vertebrates -- 8. Summary -- References -- Cardiac form, function and physiology / Anthony P. Farrell and Frank Smith -- 1. Introduction -- 2. Cardiac form and function -- 3. Cardiac physiology -- 4. Heart rate and its control -- 5. Cardiac stroke volume and its control -- 6. Coronary blood flow and its control -- 7. Summary -- References -- Hormonal and autacoid control of cardiac function / Sandra Imbrogno and Maria C. Cerra -- 1. Introduction -- 2. Catecholamines: basal control, stress, and cardiotoxicity -- 3. Angiotensin II -- 4. Natriuretic peptides: interface between myocardial performance and ion/fluid balance -- 5. Chromogranin A-derived peptides as cardiac stabilizers -- 6. Gasotransmitters as cardiac modulators -- 7. Integrated cardiac humoral signaling: the "Knot" of the NOS-NO system -- 8. Conclusions -- References -- Cardiac energy metabolism / Kenneth J. Rodnick and Hans Gesser -- 1. Introduction -- 2. Cardiac energy state and fundamentals of cellular energy metabolism -- 3. Coupling between cellular production and consumption of ATP -- 4. Energy demands of cardiac performance and homeostasis -- 5. Energy substrates and systems used to regenerate ATP -- 6. Hypoxia -- 7. Cold temperature -- 8. Body size and sex differences in cardiac energy metabolism -- 9. Remaining questions, challenges, and future directions --

	References -- Form, function and control of the vasculature / Erik Sandblom and Albin Gräns -- 1. Introduction -- 2. Gross anatomy of the vascular system -- 3. The arterial vasculature -- 4. The branchial vasculature -- 5. The venous vasculature -- Acknowledgments -- References -- Index -- Other volumes in the Fish physiology series.
Subjects	Fishes--Cardiovascular system. Fishes--Physiology. Fishes. Fishes--Cardiovascular system. Fishes--Physiology.
Form/Genre	Textbooks.
Notes	Includes bibliographical references and index.
Additional formats	Online version: Gamperl, A. Kurt. Cardiovascular system. Cambridge, MA: Academic Press, 2017 9780128041666 (OCoLC)1001979332
Series	Fish physiology series, 1546-5098; volume 36A

The synapse: structure and function

LCCN	2013039439
Type of material	Book
Uniform title	Synapse (Pickel)
Main title	The synapse: structure and function / edited by Virginia Pickel and Menahem Segal.
Edition	First edition.
Published/Produced	Amsterdam; Boston: Elsevier/Academic Press, 2014.
Description	xiv, 513 pages: illustrations; 24 cm
ISBN	9780124186750
LC classification	QP364 .S96 2014
Related names	Pickel, Virginia, 1943- editor of compilation.

Segal, Menahem, 1944- editor of compilation.

Contents Structure and complexity of the synapse and dendritic spine / Stewart -- The molecular mechanisms underlying synaptic transmission: a view of the presynaptic terminal / Ashery -- The first hour in the life of a synapse: contact formation, partner selection and onset of function / Cheyne -- Structural and functional organization of the postsynaptic density-Verpelli -- The tripartite synapse: a role for glial cells in modulating synaptic transmission / Blutstein -- Local protein synthesis at synapses / Steward -- Estrogen effects on hippocampal synapses / Milner -- Trafficking of glutamate receptors and associated proteins in synaptic plasticity / Horak -- Structural alterations of synapses in psychiatric and neurodegenerative disorders / Penzes -- Synaptic correlates of aging and cognitive decline / Hara -- Activity-induced fine structural changes of synapses in the mammalian central nervous system / Tao-Cheng -- Activity-mediated structural plasticity of dendritic spines / Muller -- Experience-dependent synaptic plasticity in the developing cerebral cortex / Aoki -- Asynaptic and synaptic innervation by acetylcholine neurons of the central nervous system / Descarries -- Prefrontal cortical dopamine transmission: ultrastructural studies and their functional implications / Sesak.

Subjects Synapses--physiology.

Synaptic Transmission.

Notes Includes bibliographical references and index.

Series Neuroscience-Net Reference Book Series

612.8

Toxins and biologically active compounds from microalgae. Volume 2, Biological effects and risk management

LCCN	2014012959
Type of material	Book
Main title	Toxins and biologically active compounds from microalgae. Volume 2, Biological effects and risk management / editor: Gian Paolo Rossini.
Published/Produced	Boca Raton, FL: CRC Press, Taylor & Francis Group, 2014.
ISBN	9781482231465 (hardcover: alk. paper)
LC classification	QK568.T67 T6952 2014
Related titles	Biological effects and risk management.
Related names	Rossini, Gian Paolo.
Contents	The mechanism of action of microalgal toxins interacting with NaV and KV channels / Kopljar, I., Peigneur, S., Snyders, D.J., and Tytgat, J -- Pharmacological actions of palytoxin / Wu, C.H. -- Molecular mechanisms of maitotoxin action / Schilling, W.P. -- The mechanisms of action of domoic acid: from pathology to physiology / Novelli, A., Fernández-Sánchez, M.T., Pérez-Gómez, A., Cabrera-García, D., Lipsky, R.H., Marini, A.M., and Salas-Puig, J. -- Cyclic imine neurotoxins acting on muscarinic and nicotinic acetylcholine receptors / Molgó, J., Aráoz, R., Iorga, B.I., Benoit, E., and Zakarian, A. -- Molecular bases of effects of azaspiracids and yessotoxins / Rossini, G.P. and Sala, G.L. -- Proteomic tools to elucidate the molecular action of micro-algal toxins / Fladmark, K.E. -- Domoic acid: biological effects and health implications / Pulido, O.M. -- Toxicity of okadaic acid/dinophysistoxins and microcystins on biological systems / Fessard, V. -- Toxicity of

	cyclic imines / Munday, R. -- Clinical applications of paralytic shellfish poisoning toxins / Lagos, N. -- Toxicology of ciguatoxins / Vetter, B.I. and Lewis, R.J. -- Toxicity of palytoxins: from cellular to organism level responses / Deeds, J.R. -- Effects of toxic microalgae on marine organisms / Landsberg, J.H., Lefebvre, K.A., and Flewelling, L.J. -- Phycotoxins: seafood contamination, detoxification and processing / Lassus, P., Bourdeau, P., Marcaillou, C., and Soudant, P. -- Coupled nature-human (CNH) systems: generic aspects of human interactions with blooms of Florida red tide (Karenia brevis) and implications for policy responses / Hoagland, P. -- Modeling of harmful algal blooms: advances in the last decade / Francks, P.j.S. -- Challenging times for the detection of marine biotoxins in the EU / Gago-Martínez, A. and Braña-Magdalena, A. -- Risk management of marine algal toxins in China / Wang, J., Huang, H., and Wu, J. -- Some models of risk management posed by toxic microalgae and microalgal toxins in Africa / Taleb, H. and Foord, C.J. -- International initiatives to assess and manage the risk of biotoxins in bivalve molluscs / Ababouch, L.
Subjects	Toxic marine algae. Microalgae
Notes	Includes bibliographical references and index.

RELATED NOVA PUBLICATIONS

DIFFERENTIAL ROLES OF THE α4B2 AND A7 NICOTINIC ACETYLCHOLINE RECEPTORS IN NICOTINE ADDICTION: IMPLICATIONS FOR DEVELOPING PHARMACOTHERAPIES FOR SMOKING CESSATION[*]

Xiu Liu[†] and Yongzhen Gong

Department of Pathology, University of Mississippi Medical Center, Jackson, MS, US

Nicotine, the addictive component of tobacco products, exerts its pharmacological actions via activating the nicotinic acetylcholine receptors (nAChRs). Among an increasing number of nAChR subtypes, the α4β2- and α7-containing nAChRs are the two major ones, accounting for about 95% of the whole nAChR population in brain. These receptor subtypes

[*] The full version of this chapter can be found in *Advances in Medicine and Biology. Volume 102*, edited by Leon V. Berhardt, published by Nova Science Publishers, Inc, New York, 2016.

[†] Corresponding Author: Xiu Liu, MD, PhD, Associate Professor, Department of Pathology, University of Mississippi Medical Center, 2500 North State Street Jackson, MS 39216, Phone: (601) 984-2875, Fax: (601) 984-1531, Email: xliu@umc.edu.

show considerable differences in their distribution and functions. The research work over the past decade or so in our laboratory has demonstrated differential roles of the α4β2 and α7 nAChRs in the process of nicotine addiction. Specifically, rat models of nicotine consumption and cue-induced relapse were used to examine the effects of selective antagonism of these two nAChR subtypes on the primary reinforcement of nicotine and the conditioned reinforcing actions of nicotine-associated environmental stimuli (cues). Male Sprague-Dawley rats were trained to intravenously self-administer nicotine (0.03 mg/kg/infusion, free base) on a FR5 schedule of reinforcement and nicotine-conditioned cues were established via association of a sensory stimulus (5-s tone/20-s lever light on) with each nicotine infusion. After extinguishing rats' lever-pressing responses by withholding nicotine and its cues, the reinstatement tests were conducted where the nicotine cues were re-presented response-contingently without nicotine availability. Under both the self-administration and the reinstatement test conditions, prior to the test sessions, rats were subjected to systemic administration of either a α4β2-selective antagonist dihydro-β-erythroidine (DHβE) or a α7-selective antagonist methyllycaconitine (MLA). In the self-administration test sessions, DHβE but not MLA reduced lever-pressing responses and hence suppressed nicotine self-administration. In contrast, in the reinstatement test sessions, MLA but not DHβE effectively decreased the magnitude of responses on the previously active and nicotine-reinforced lever that were reinstated by response-contingent representations of nicotine cues. These findings indicate that cholinergic neurotransmission via activation of the α4β2 but not α7 subtype of nAChRs participates in the primary reinforcing actions of nicotine, whereas, activation of the α7 rather than α4β2 subtype of nAChRs plays a role in the mediation of the conditioned incentive properties of nicotine cues. Therefore, this work supports the continued effort to develop cholinergic agents aiming at the α4β2 nAChRs for reducing or stopping smoking. However, it is suggested that manipulation of α7 nAChR activity may prove to be a promising target for the development of pharmacotherapies for the prevention of smoking relapse triggered by exposure to environmental cues.

Development of Nicotinic Acetylcholine Receptor Imaging Probes and Their Use for the Functional Analysis of Nicotinic Acetylcholine Receptors in Neuropathic Pain*

Masashi Ueda[1] and Hideo Saji[2,†]
[1]Graduate School of Medicine, Dentistry, and Pharmaceutical Science, Okayama University, Okayama, Japan
[2]Graduate School of Pharmaceutical Sciences, Kyoto University, Kyoto, Japan

Nicotinic acetylcholine receptors (nAChRs) in the central nervous system have been implicated in not only higher brain functions such as memory, learning, and recognition, but also a variety of physiological functions such as neuroprotection and analgesia. The development of a technique for the noninvasive measurement of the distribution and density of nAChRs in the brain thus offers a promising avenue of providing valuable information for identifying the roles of nAChRs in the central nervous system. Recent years have seen great interest in the field of "molecular imaging" for understanding the action of molecules generated in vivo by analyzing changes in gene and protein expression and the resulting changes in cellular and tissue function. One such technique is nuclear medical molecular imaging, which has the advantages of being noninvasive and highly sensitive. To perform the nuclear medical molecular imaging, a radioactive probe is necessary; we have successfully developed a radiolabeled imaging probe targeting nAChRs and successfully visualized and quantified nAChRs in vivo. Moreover, we have

* The full version of this chapter can be found in *Horizons in Neuroscience Research. Volume 26*, edited by Andres Costa and Eugenio Villalba, published by Nova Science Publishers, Inc, New York, 2016.

† Corresponding Author's E-mail: hsaji@pharm.kyoto-u.ac.jp.

identified novel expression sites of nAChRs that are implicated in analgesic activity in the brains of a rat model of neuropathic pain by using this imaging technique. We designed a novel nAChR agonist and demonstrated its efficacy by affecting nAChRs expressed in the novel sites. These findings indicate that nAChRs expressed in these sites may be potential targets for the development of novel analgesics. The experimental details and results are described in this chapter.

INDEX

A

B

C

D

E

F

G

H

I

T

U

V

W

α

FROGS: BIOLOGY, ECOLOGY AND USES

EDITOR: James L. Murray

SERIES: Animal Science, Issues and Professions

BOOK DESCRIPTION: In this book, the authors gather and present topical research from across the globe in the study of the biology, ecology and uses of frogs.

HARDCOVER ISBN: 978-1-61324-667-2
RETAIL PRICE: $225

FROGS: GENETIC DIVERSITY, NEURAL DEVELOPMENT AND ECOLOGICAL IMPLICATIONS

EDITOR: Henry Lambert

SERIES: Animal Science, Issues and Professions

BOOK DESCRIPTION: In this book the authors present current research in the study of frogs. Frog`s neuromuscular junction (NMJ) is a classic and favorite object which have played a leading role in developing understanding of the basic mechanisms of synaptic transmission and secretion of neuromediator.

HARDCOVER ISBN: 978-1-63117-626-5
RETAIL PRICE: $230